Table of Contents

Preface

Training Purpose: This curriculum was directly inspired by the US Marine Corps' Combat Hunter program. Created in 2007, in response to a dramatic increase in precision fire causalities in Baghdad, Combat Hunter is systematic training designed to improve cognitive skills, showing personnel how to read the human terrain, establish a baseline, detect an anomaly, and make decisions "left of bang." In other words, Combat Hunter was designed to train personnel to anticipate danger and meet it proactively. In an irregular conflict, this enables personnel to be the "hunters"—not the "hunted."

CODIAC integrates the USMC Combat Hunter principles, along with proven battlefield decision-making and irregular warfare instruction from across the Joint services. The goal of CODIAC is to enhance the ability of individuals and small teams to address irregular challenges by training enhanced observation, battlefield sensemaking, human terrain pattern recognition, and environmental analysis (including knowledge of combat tracking).

Training Location: In unit (home station)

Course Length: This course should require approximately 200 hours to complete. A sample schedule is provided; however, units should develop schedules that meet their own goals.

Collaboration: This curriculum incorporates the latest educational theories regarding dynamic decision-making, human social cultural behavior analysis, and battlefield sensemaking. Military and academic personnel, cognitive psychologists and training scientists from industry, and the USMC Combat Hunter subject-matter experts collaborated in the creation of this curriculum. Development of this curriculum was sponsored by Joint Forces Command (USJFCOM) in collaboration with Joint Task Force North (JTF–N).

In its role as the joint integrator of training, USJFCOM's contribution to this effort was to collaborate with the Services and other partners to find the best-of-breed ideas, build upon them, and bring them to the attention of the Defense community in a way that reinforces the Services best products, insights, and capabilities on a Joint "shelf."

Target Audience of this Manual: This Program of Instruction (POI) was designed for military personnel in the Joint Defense community, including the Army, Navy, Air Force, and Marine Corps. Further, to meet Department of Defense objectives, the POI may be used for training interagency paramilitary personnel, such as Border Patrol or Police Officers, as well as multinational allies.

The POI was designed as a high-level "teacher's guide" for qualified trainers to use to train novice personnel on enhanced observation, tracking, and combat profiling. This instructional manual was designed to help trainers convey dynamic decision-making skills in an accessible and innovative, yet straightforward and systematic, manner.

Using this POI: This POI was designed for use by individual units, and it is intended to be used at their discretion. Ideally, CODIAC trainers will have completed a train-the-trainer CODIAC, Combat Hunter, or similar course. This POI will benefit all combat personnel, particularly those deploying to an active area of operations. Instructional delivery and content guidance are offered for each module. However, this guide is not a complete source of all CODIAC subject matter. Additional resources are recommended throughout.

Subject Matter Experts

Greg Williams

On a typical day, one might encounter an outgoing individual who seems to never stop. He is in constant motion and seeking, processing, and associating information he needs to interpret the intricacies of the patterns that immerse us all. Greg Williams is that individual. Greg was reared in a survival situation in a large urban area. He dodged in and out of trouble, a natural fighter who honed his skills in numerous encounters where gang ethics and honor were at stake. Early on he became aware of the subtle indicators that survivors seem to notice... symbols, weapons, body language, baselines.

He survived those early street encounters, which included numerous contacts with law enforcement. Eventually, circumstances required Greg to enter the military where he served as an infantryman. He moved on to a career in law enforcement, where he gained a reputation for sensing out criminals and using his street-sense to combat crime.

Greg Williams is a transformational character, who adapted his skills and abilities to defeat insurgents on the battlefields of Asia. As a trainer and practitioner of the art and science of profiling, Greg found meaning around him. He adapted his skills as a detective to the requirements of today's infantryman. He proposed a new lexicon and added a dimension to problem solving that gave the user the advantage in time and space to kill, capture, or contact an adversary by combining his precepts about human nature with his insights about alien cultures and his expert knowledge of weapons and warfare.

↘ RESOURCE:
Resource DVD, 0000.1 – Greg Williams Interview Video

Why should Greg Williams be considered a subject matter expert? He is a master of the skills, and he has integrated them into a new combat capability—human terrain pattern recognition. Greg's work goes beyond fluency. Many individuals possess the basic ability of reading the intentions of others, but Greg's performance goes beyond. He has demonstrated an ability to re-calibrate his understanding and efficiently adapt his mental model to emerging situations. He has achieved a level of performance that is characterized by originality and creativity. He has effectively integrated his life experiences with understanding of human behavior to amplify the performance of small units who operate on the "edge of chaos." It is a new frontier that enables Greg and his protégés to interact and relate to their operating environments. In this sense, he has captivated the imagination and spirit of warfighters by making them the hunter instead of the hunted in survival situations.

David Scott-Donelan

As we scan the environment for others with prodigious abilities as sense makers and difference makers, we encounter David Scott-Donelan. A renowned human tracker whose has extensive experience in counter-guerilla warfare, David is an outlier. He is the product of a military experience, where he served and led various regular and unconventional units such as the Seleous Scouts and the Special Air Services (SAS). As a young lieutenant in the Rhodesian Army, he acquired the tactical savvy and resilience while fighting active insurgencies throughout Africa. He demonstrated an uncanny ability to get in the mind of his quarry and to track them relentlessly. He is a natural leader who inspires, teaches and bleeds with his followers. To those who see him today, David stands out as an individual who lives and breathes the ethos of the warrior class.

David is vital and fit for someone who has passed his 70th birthday. He is a lifelong learner and a passionate reader who consumes a steady diet of nonfiction materials. He is also a student of human nature, which he believes gives him the advantage when tracking or analyzing evidence. Today, he is surrounded by a cadre of protégés who practice the craft of tracking. He is an effective storyteller, who integrates his lived experiences with lesson materials to inform listeners on how and why tracking works. He and his team are painstaking attention-to-detail men. Like David, they are intellectually curious, always trying to explain why things happen.

When asked, he speaks proudly about his military roots. David also reveals that he comes from a long line of professional soldiers. This lineage begins with a great grandfather who spent 57 years in the service, which culminated in serving as the Yeoman of the Guard for Queen Victoria. David's service spanned nearly three decades in several war zones. His expertise as a tracker was refined throughout that period.

Why should David Scott-Donelan be considered an outlier? For nearly 30 years, David has applied the tactical, technical and decision making skills of a tracker. Perhaps more importantly, he has mentored hundreds of others as role model and coach. His protégés are throughout law enforcement and military organizations. David is a lateral thinker, who possesses a natural ability to interpret human spoor and apply his understanding in high risk problem settings. He is inner directed, which gives him the ability to retain details and focus on tracking tactics and techniques. As a combat tracker, David was invariably dealing with dangerous situations where he was required to remain vigilant and aware for long periods of time. This level of mental stamina contributes significantly to his judgment and decision making while tracking.

➘ RESOURCE:
Resource DVD, 0000.2 – David Scott-Donelan Interview Video

Training Outcomes

This program intends to teach the following:

OBSERVE

Use of enhanced observation techniques

1. Using organic assets to make positive identifications
2. Making innovative use of optics to help construct a baseline or profile
3. Shifting field of view to avoid focus lock
4. Efficient use of refocusing in visual scans to include both near and far objects in the scene
5. Making effective and efficient identification of anchor points and indications of anti-tracking
6. Making effective and efficient identification of habitual areas and action indicators
7. Effortlessly using observation techniques that do not require conscious attention

Identification of critical event indicators

8. Establishing a baseline to extract normalcy
9. Looking for anomalies outside of the baseline
10. Looking for signature behaviors (e.g., of a high-value target) via a cluster of cues
11. Looking for signature locations (e.g., habitual areas) through a cluster of cues

ORIENT

Interpretation of human behavior cues

12. Taking someone else's perspective
13. Effectively and efficiently identifying leaders
14. Orienting observation or tracking toward potentially hostile players and ignoring neutrals
15. Working with others to construct a behavior profile of a person, event, or quarry

Synthesis of ambiguous information

16. Inducing a pattern from a few individual cues
17. Generating explanatory storylines that tie individual items of information together
18. Imagining alternative courses of action or event outcomes by what-if mental simulations
19. Anticipating what will happen next

DECIDE

Proactive analysis and dynamic decision-making

20. Looking for prototypes to guide rapid recognition and decision-making
21. Detecting an unfolding event or activity by identifying a piece of it and inferring the rest
22. Using appropriate criteria (e.g., three cues) to make timely but accurate decisions
23. Taking an evidence-based approach, using hard data to confirm or disconfirm hypotheses
24. Not settling for unexplained events or evidence but looking for antecedents to a situation
25. Making effective decisions in spite of high stress conditions

MAINTAIN

Employment of cognitive discipline

26. Using tactical patience to avoid committing too soon or going to kinetics unnecessarily
27. Using geometry of fires to create an interlocking network of optics, intel, and communications
28. Keeping an open mind to the unexpected (recognizing that there are unknown variables)
29. Recognizing when one's situational awareness is low and how to mitigate the condition
30. Trusting that one's skills will overcome obstacles in the difficult situations
31. Developing an internal sense of time in order to know when a situational judgment needs to be updated
32. Employing stress reduction strategies to manage physiological stress reactions
33. Recognizing when stress is affecting other team members' actions and helping them refocus their attention

Unit 1: Introduction to CODIAC

1

This course, called Combat Observation and Decision-making in Irregular and Ambiguous Conflicts (CODIAC), is intended to help personnel, at all ranks and in various occupational specialities, develop sophisticated decision-making abilities for complex, irregular, and ambiguous environments. This unit will explain the importance of CODIAC skills from the irregular warfare perspective. Trainees will learn the purpose of CODIAC, and they will see how such skills can benefit the range of combat personnel.

Suggested Prerequisites:

- None

Terminal Learning Objective:

- As an introductory unit, this section serves as a primer to all six of the major training objectives. At the conclusion of this unit, trainees will be able to describe the purpose of CODIAC and explain how CODIAC skills can enhance their individual role in an operational area.

Enabling Learning Objectives:

- Describe the consequences of being reactive (versus proactive) on the battlefield
- Describe the benefits of making decisions "left-of-bang"
- Evaluate decisions in various examples from a legal/moral/ethical perspective

Estimated Time Allotted for Instruction: About 8 hours

Speciality Support Requirements:

- No special facilities required

1001: Course Introduction

Introductory
Lecture and discussion
About 60 minutes

Terminal Objective: This module introduces trainees to the goals and objectives of the course. Trainees will learn about critical incidents in which CODIAC skills were used, or could have been used, to prevent serious consequences. Trainees will also learn about the material to be covered, and they will be introduced, via anecdotes, to real-world uses of combat profiling and combat tracking.

Module Vocabulary:

- CODIAC
- Irregular Warfare
- Left-of-Bang

Resource DVD:

- 1001.1 – Intro, Video
- 1001.2 – Juba Sniper, Video
- 1001.3 – Left-of-Bang, PPT
- 1001.4 – Combat Hunter in *USMC Gazette*, PDF

Instructor Activities:

1. Review the Intro video with the class
2. Discuss "left-of-bang" and explain its importance
3. Lead a discussion on sensing danger and responding to incidents
4. Discuss the Juba sniper and creation of the Combat Hunter course
5. Discuss the CODIAC unit topics and objectives

Synopsis of Module Topics:

Introduction: From 2005–2007, a man (or possibly a group) called the "Juba Sniper" terrorized American warfighters in Baghdad. In a series of Internet-published propaganda videos, Juba can be seen killing American warfighters. In one of his videos, released October 2006, Juba claims to have killed 645 US Soldiers and Marines. Regardless of whether these deaths were truly Juba's handiwork, the US casualty report verifies a dramatic increase in precision fire causalities during this time period. Marines and Soldiers were being ***hunted***.

Juba Sniper: The Juba sniper example shows how poor situational awareness, a lack of combat observation skills, and ignorance of how to "read" the human and physical terrain can be deadly.

USMC Combat Hunter: In January 2007, the US Marine Corps (USMC) sought a novel solution to the Juba sniper and similar irregular warfare challenges. With assistance from the Marine Corps Warfighting Lab (MCWL), a diverse group of subject matter experts were assembled with a common goal: Turn Marines from the ***hunted*** into the ***hunters***.

The USMC realized that certain people were better able to detect snipers and improvised explosives. After examining their backgrounds, the Marines realized that the most successful "battlefield hunters" were those who could read the environment: the physical and/or social landscape. Combat Hunter was designed to train these skills. The program focuses on enhanced observation, combat tracking (i.e., reading the physical terrain), and combat profiling (i.e., reading the human terrain).

Border Hunter: Because the Combat Hunter skill set appeared to have transferability to a wider community, in April 2010 Joint Task Force North (JTF-N) held a special 20-day "Border Hunter" version of the Combat Hunter course. It was delivered to 43 trainees from the Army and Law Enforcement Agencies, including 18 Border Patrol personnel (hence the name). JTF-N redesign Range Golf at Ft. Bliss to specifically accommodate the Border Hunter training scenarios, and they hired the original Subject Matter Experts to teach the course. This manual is based upon the instruction offered during that Border Hunter event.

CODIAC: The Combat Observation and Decision-making in Irregular and Ambiguous Conflicts (CODIAC) initiative is an effort to distill the core competencies of Combat Hunter into a format usable across the Joint Services.

Left-of-Bang: CODIAC principles help personnel observe, analyze, and decide before the enemy acts, left-of-bang. CODIAC skills also help personnel learn the indicators to look for after an incident occurs, so that they can prevent the next occurrence—that is, acting before the *next* bang.

Training Objectives: CODIAC includes six major training objectives:

1. *Use of enhanced observation techniques:* The knowledge, skills, and attitudes (KSAs) associated with this objective involve understanding the physiological and cognitive processes of perception, the limits of these processes, and how to overcome the limitations.
2. *Identification of critical event indicators:* The KSAs associated with this training objective involve interpretation of observed events. In other words, this objective highlights the need to effectively analyze a scene in order to make sense of a battlespace.
3. *Interpretation of human behavior cues:* As an extension of the previous training objective, this objective also focuses on analysis of observations but emphasizes the ability to interpret human indicators.
4. *Synthesis of ambiguous information:* These KSAs focus on the development of synthesis and prediction abilities.
5. *Proactive analysis and dynamic decision-making:* Once cues have been detected, interpreted, and synthesized, a decision must be made. This training objective involves KSAs associated with making effective decisions despite having only partial or ambiguous information.
6. *Employment of cognitive discipline:* Finally, this training objective focuses on maintaining peak performance by being aware of, and then mitigating, limitations (such as fatigue or stress) that may impair one's ability to make effective observations and/or decisions.

Discussion Questions:

1. Have you ever experienced a "sixth sense" that warned you about impending danger? What was that sensation? Why did you have it?
2. Are commanders/supervisors likely to give "sixth sense" much credence? Why or why not? What would make it more believable?
3. What happens after a critical incident? How do you track down your quarry? How do you figure out who was responsible?

"THEY HAVE NEVER SEEN JUBA. They hear him, but by then it's too late: a shot rings out and another US soldier slumps dead or wounded. There is never a follow-up shot, never a chance for US forces to identify the origin, to make the hunter the hunted. He fires once and vanishes"

—*Rory Carroll*

REMEMBER!

Today's battlefield is constantly changing and requires a unique ability to deal with the civilian populace, decentralized command and control, and the authority and consequences of individual actions—all while fighting an adaptive enemy.

1002: Joint Distributed Operations

Introductory
Lecture and discussion
About 60 minutes

Terminal Objective: This module introduces trainees to the concept of distributed operations and the challenges of irregular warfare that necessitate a distributed strategy. At the conclusion of this module, trainees will be able to list several challenges of irregular warfare and identify how different Services are overcoming these challenging by placing more emphasis on small, dismounted units.

Module Vocabulary:

- Counterinsurgency
- Distributed Operations
- Full-Spectrum Operations

Resource DVD:

- 1002.1 – *Learning Counterinsurgency*, PDF
- 1002.2 – *NDS 2005*, PDF
- 1002.3 – *ECO Doctrine*, PDF
- 1002.4 – *Army Field Manual 3-0*, pages 47–68, PDF

Instructor Activities:

1. Ask trainees to read the recommended resources
2. As a class, interactively discuss irregular warfare
3. Discuss its implications, such as distributed operations
4. Discuss how CODIAC skills can assist in irregular conflicts

Synopsis of Module Topics:

The Challenge: Military and law enforcement personnel confront a range of irregular conflicts, fought by adversaries who are fierce, agile, unpredictable, and willing to use extreme methods of violence. To confront these challenges, US forces and civil authorities are adopting distributed strategies of operations; these strategies place greater emphasis on small, dismounted teams. The individuals and junior leaders who comprise these teams are responsible for making rapid, difficult decisions. They also have greater responsibility for intelligence collection, surveillance, and communications. Plus, the consequences of failure can be substantial, having strategic implications beyond the immediate tactical operations.

Irregular Warfare: Irregular warfare is characterized as a violent struggle among state and non-state actors for legitimacy and influence over the relevant population(s). Irregular warfare combatants often favor indirect and asymmetric approaches, though they may employ the full range of military and other capacities, in order erode an adversary's power, influence, and will.

Strategic Corporals: Key decision-making responsibilities are now often pushed to more junior personnel. To act effectively these personnel must possess numerous enabling competencies, including critical thinking abilities, situational awareness, tactical patience, cultural attentiveness, ethical/moral solidity, and an adaptive stance, and the confidence to use their skills within the complex, ambiguous environment of the modern battlespace. In his article on "Learning Counterinsurgency," General Petraeus discusses 14 lessons-learned from the recent conflicts in Iraq and Afghanistan. Although his article focuses on the Middle East, the General's words provide insight for a range of military and law enforcement operations. One of his most poignant recommendations is #12:

> Observation Number 12 is the admonition to *remember the strategic corporals and strategic lieutenants*, the relatively junior commissioned or noncommissioned officers who often have to make huge decisions, sometimes with life-or-death as well as strategic consequences, in the blink of an eye. ...Do everything possible to train them before deployment for the various situations they will face, particularly for the most challenging and ambiguous ones (page 9).

Distributing Operations: The collective aim of distributed operations is to meet the challenges of irregular warfare by unleashing the power of small units, enabling personnel to operate autonomously at increasingly lower echelons, disaggregate against unconventional threats, quickly aggregate against more conventional threats, and easily integrate Joint capabilities, such as intelligence, surveillance, and reconnaissance (ISR) and Joint fires. The intent is not to make regular forces into special forces but to recognize that the individual, the leader, and the small unit are critical players on an increasingly decentralized and dispersed battlefield.

...WE REQUIRE the capabilities to identify, locate, track, and engage individual enemies and their networks. Doing so will require greater capabilities across a range of areas, particularly intelligence, surveillance, and communications.

—*National Defense Strategy (NDS) 2005, page 15*

Full-Spectrum Operations: Another challenge of the current operational environment is that US military forces and civil authorities are asked to conduct a wider range of operations than ever before. The Army and Marine Corps refer to this as "full-spectrum operations." Full-spectrum operations combine offense, defense, and stability/support operations. The implications of this effort effect all levels. For instance, commanders and strategic planners must develop greater understanding of the social, cultural, and psychological motivations of worldwide populations, and they must develop wider-ranging strategies that consider the second-, third-, and fourth-order effects of their actions. Meanwhile, relatively lower-ranked personnel must also develop significantly expanded competencies, beyond their military occupational specialties, to support the range of kinetic (i.e., combat) and non-kinetic (i.e., non-combat) missions they will be asked to carry out.

Implications: The reality of irregular warfare, counterinsurgency, distributed operations, and full-spectrum operations leads to several implications. First, small units (i.e., squads, platoons, and companies) now have greater responsibly for intelligence, surveillance, and communications tasks. Second, small unit leaders now have greater empowerment to make critical battlefield decisions—not just at the tactical level but also with strategic consequences. Third, all personnel are tasked with a greater range of missions, from reconstruction duties to force-on-force armed conflict. Finally, to rise to these challenges, small unit personnel and leaders must develop greater higher-order cognitive competencies, broader knowledge of the tactical strategic battlespace, and intuitive decision-making abilities for use in ambiguous environments. This CODIAC manual was designed to partially address these training needs.

Discussion Question:

1. What elements of irregular conflicts do you face in your operational environment? How does your unit address these challenges—both formally and informally?

1003: The Incident Timeline

Introductory
Lecture and discussion
About 60 minutes

Terminal Objective: This module describes the incident timeline, as well as how CODIAC skills can be applied before and after the critical incident. Upon completion of this module, trainees will be able to describe how CODIAC skills can be used left-of-bang and right-of-bang. They will also be able to list several examples of CODIAC skills applied to each of these stages of the timeline.

Module Vocabulary:

- Pre-Event Indicators
- Right-of-Bang
- Rule of Three

Resource DVD:

- 1003.1 – Timeline, Video

KEY SKILL

Anticipating what will happen next

Instructors should emphasize the importance of acting proactively, that is, left-of-bang. Discuss specific critical events that the trainees have experienced and attempt to determine what pre-event indicators were present in each case.

Instructor Activities:

1. Define the incident timeline; use the provided Timeline video
2. Remind trainees of the importance of acting left-of-bang
3. Define and discuss *pre-event indicators*
4. Discuss the *rule of three*
5. Discuss the use of CODIAC skills right-of-bang

Synopsis of Module Topics:

Introduction: The incident timeline describes the activities, in time order, that led up to, occurred during, and then happened after a critical incident (i.e., a "bang"). Personnel can use CODIAC skills left-of-bang and right-of-bang.

Incident Timeline: "Left-of-bang" refers to all actions that occur before an incident; these may include pre-event indicators, actions to interrupt the incident, or reconnaissance of an objective. Actions "left-of-bang" are *proactive*. "Right-of-bang" refers to all reactions that occur after an incident. Actions "right-of-bang" are *reactive*. Personnel should strive to be proactive, identifying and mitigating threats before they occur.

Left-of-Bang: As discussed earlier in this unit, a primary goal of CODIAC training is to help personnel make accurate decisions left-of-bang, before a critical incident occurs. In order to act left-of-bang, a person must be able to observe and understand the *pre-event indicators* that suggest that a critical incident is about to occur.

- *Rule of Three:* Keep in mind that, in most cases, a single cue is not enough to make a decision, unless that cue is substantial (e.g., an immediate threat to a person demanding that he/she act in self-defense). However, once three anomalies have been detected, a decision must be made.

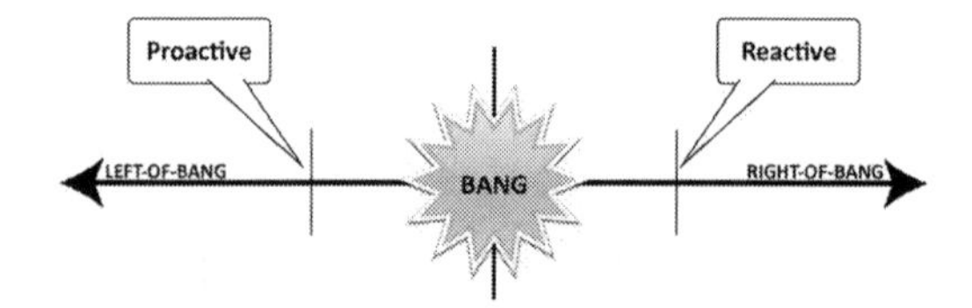

Examples of what personnel can do, left-of-bang, include:

- Create a baseline of what's "normal" for an area
- Observe suspects to establish their typical tactics and procedures
- Scout out potential attack locations
- Detect anomalies from the baseline
- Identify behaviors that are out of place
- Identify suspicious environmental signs (e.g., tracks at a location)

Right-of-Bang: Warfighters and law enforcement personnel can use CODIAC skills to analyze critical events after they occur, identifying indicators to prevent the next critical incident—in other words, acting left-of-*the-next*-bang. For example, CODIAC skills can help personnel track down insurgents and fugitives or identify terrorist and criminal networks. To do this, personnel may need to exercise tactical patience. For example, a police officer who observes a drug dealer on the street could arrest that person, or the police officer could follow the drug dealer until he makes contact with his supplier. Then the police officer could bring down the drug network instead of just the dealer.

Examples of what personnel can do, right-of-bang, include:

- Identify tracks leading away from a scene
- Identify behavioral anomalies of nearby people
- Identify environmental effects, such as odd crowd reactions
- Analyze a site for clues to the enemy's tactics or motivations
- Analyze the incident for clues to the network involved

"WE'VE FOUND IN MOST of these cases of targeted violence there has been some pre-event indicator, either behavior, comments or action on the part of people that, in retrospect, was an indicator that something was going on that might have born [*sic*] more investigation."

—*Gene Ferrara, director of public safety and police chief for the University of Cincinnati in response to the Virginia Tech massacre (quoted from an article by Leischen Stelterin in "Security Director News")*

Discussion Questions:

1. Think of a critical event that you, personally, experienced. What pre-event indicators can you identify in retrospect? Where there any behavioral cues (e.g., from the other people involved) that might have been pre-event indicators?
2. Again, thinking of real events, what sort of post-event indicators were available that could (or were) used to prevent another critical incident? What sort of post-event indicators may be most useful for your operations, in general? What key information should *you* (personally) look for, immediately following an incident.
3. What abilities does someone need in order to act left-of-bang? They need to be able to *observe* the pre-event indicators, *orient* to their meaning, make effective *decisions* about what actions to take, and then take those *actions* that circumvent the "bang" (or critical event).

1004: Decision-Making Cycle

Introductory
Lecture and discussion
About 60 minutes

Terminal Objective: This module introduces the OODA-Loop and the three primary CODIAC action options. Upon completion of this module, trainees will be able to accurately define and describe the simplified OODA-Loop. They will be able to list and describe the three CODIAC action options, and they will be able to argue why accelerating the speed of decision-making cycle is critical for mission success.

Module Vocabulary:

- OODA-Loop

Resource DVD:

- 1004.1 – OODA, Video

"No one, not even Karl Von Clausewitz, Henri de Jomini, Sun Tzu, or any of the past masters of military theory, shed as much light on the mental and moral aspects of conflict as Boyd."

—*James Burton (1993)*

Instructor Activities:

1. Describe the importance of decision-making in irregular conflicts
2. Review the OODA video
3. Define and discuss Boyd's OODA-Loop decision cycle
4. Discuss, in detail, the three CODIAC action options
5. Discuss the importance of accelerating the OODA-Loop
6. Explain that CODIAC training is about improving and accelerating the decision-making cycle

Synopsis of Module Topics:

Introduction: Central to success in irregular and ambiguous conflicts are small-unit personnel and leaders who can think critically, make sound decision, and maintain high awareness when faced with dynamic situations. To master these capabilities personnel must understand the decision-making cycle.

Boyd's Decision Cycle: The decision-making cycle is a constantly revolving process that takes place in the mind every second of every day, with regard to all tasks, from mundane to complicated. This cycle follows the pattern of Observe-Orient-Decide-Act (OODA). Created by USAF Colonel John Boyd, the OODA cycle describes how the mind deals with its outside environment and translates it to action.

- *Observe:* Observation, the first step in the OODA-Loop, is a search for information relevant to the tactical situation. This could include anything, from the enemy's tactics to the moral context. Observation is not passive step. It requires an active effort to seek out all the available information by whatever means possible.
- *Orient:* Orientation involves synthesizing the observed information to form an awareness of the overall context. In other words, orientation helps turn information into understanding, and it is understanding that leads to good decisions.
- *Decide:* Decisions are based upon the perceived observations, as well as training, experience, rules of engagement, orders, and directives. Through repetitive training, some decisions can become automatic or reflexive (for example, immediate action drills for weapon malfunctions). It is the goal of CODIAC training to help decisions in irregular and ambiguous situations become more automatic.
- *Act:* The final step, act, describes the implementation of a decision. For CODIAC skills, personnel need only consider three actions: Kill, Capture, or Contact. If none of these actions are appropriate, then "let it go."

CODIAC Action Options: Personnel in the field must make very rapid decisions. To facilitate this, consider only these three actions:

- *Kill:* Choose this option when a threat is so immediate and deadly that terminating it is the only course of action that will prevent damage to oneself, fellow personnel, or civilians.
- *Capture:* If a person, object, or scene appears to possess relevant intelligence (e.g., someone is identified as a Person of Interest), then he, she, or it should be captured. Capture may literally mean apprehension and detention. However, other forms of "capture," may include photographing a suspicious person, videotaping a scene, or writing down the license plate of a vehicle of interest.
- *Contact:* If an observed individual displays certain behaviors that require further investigation, personnel may contact that person. Contact takes places in many forms and degrees. Simply maintaining observation for a longer period of time is a form of contact; as well, physically approaching and debriefing individuals to gain more information is a form of contact.
- *Leave it Alone:* All anomalies should be contacted in some form or fashion as described above. However, if none of the above actions are appropriate, then "leave it alone." Some anomalies are not worth further observation, nor should they be killed or captured. Knowing where to focus *and what to ignore* is crucial.

Speed: Speed is a weapon. This is true for the decision cycle, as well as traditional kinetic operations. The goal is to accelerate the OODA-Loop, particularly the *orientation* and *decision* phases.

Enemy Decision Cycles: Enemies also makes decisions via the OODA-Loop (although they may not consciously realize it). In conflict, whomever performs the OODA-Loop most rapidly will have a significant advantage. Hence, it is imperative that warfighters and law enforcement agents strive to execute their OODA-Loops most efficiently. If achieved, then it is the warfighter or law enforcement agent who controls the tempo of the conflict, and the enemy become the reactive participant.

Discussion Questions:

1. What phase of the OODA-Loop requires the most time to complete? Why is that phase so time consuming? Have you ever experienced a sense of "mental slowness" when you were involved in a critical event?
2. How might you increase the speed at which you *orient* during an operation? How about increasing the speed at which you *decide*?

"TIME IS THE DOMINANT parameter. The pilot who goes through the OODA cycle in the shortest time prevails because his opponent is caught responding to situations that have already changed."

—*Harry Hillaker, chief designer of the F-16*

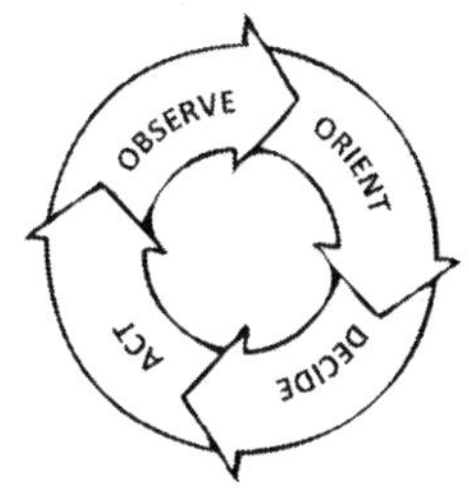

The OODA-Loop describes the process of decision-making, which is constantly unfolding in each person's mind. This diagram (and the description in this module) presents a simplified version of Boyd's theory.

KEY SKILL

Orienting observation or tracking toward potentially hostile players and ignoring neutrals

Instructors should emphasize the importance of spending mental energy only on the potentially hostile individuals in a location. If a person or group does not appear hostile, then "leave it alone," and kill, capture, or contact someone else.

1005: Legal, Moral, Ethical Decisions

Introductory
Lecture and discussion
About 210 minutes

Terminal Objective: All trainees must learn the criticality of ethical actions, and they must learn to consider their personal actions from a strategic perspective. Upon completion of the module, trainees will be able to clearly discuss the importance of ethical behavior and list specific ways in which they can help engender ethical behavior in their agency.

Module Vocabulary:

- Warrior Ethos

Resource DVD:

- 1005.1 – Ethical, Video
- 1005.2 – Ethical, PPT
- 1005.3 – *Abu Ghraib*, PDF
- 1005.4 – *Wounded Dishonored*, PDF
- 1005.5 – *Case Studies*, PDF

"...It takes one screw-up to erase 1,000 jobs done well. This is why every Marine must keep their heroes in mind when contemplating doing something illegal, offensive or dangerous."

—Joseph R. Chenelly

Instructor Activities:

1. Describe the importance of making legal, moral, ethical decisions
2. Compare and contrast the terms *legal*, *moral*, and *ethical*
3. Review the Ethical video
4. Review and discuss the Ethical PowerPoint presentation
5. Hold an interactive discussion on legal versus ethical actions
6. Hold an interactive discussion on how each trainee can better support legal, moral, and ethical behavior in his/her teams; use ethical scenarios to facilitate parts of the discussion (see resource 1005.4)

Synopsis of Module Topics:

Introduction: Personnel must make rapid decisions, left-of-bang, but this process should be tempered with an understanding that all decisions must be legal, moral, and ethical. The identification of an anomaly is not an excuse to directly reason that a target must be killed, but that it must be analyzed within its context and relevance, as well within the legal, moral, and ethical perimeters.

- *Legal:* Conforming to, or permitted by law and within the Rules of Engagement (ROE).
- *Moral:* Pertaining to, or concerned with, the principles or rules of right conduct. That is, the distinction between right and wrong.
- *Ethical:* Being in accordance with society's rules, standards, and expectations for right conduct, especially the standards of a profession; in other words, being moral as well as socially appropriate. The expectation to act ethically is placed on every warfighter or law enforcement agent who is representing America.

Legal vs. Ethical: *Legal* and *ethical* are not always the same. Sometimes a legal choice is not an ethical choice. For example, cheating on one's spouse is not illegal, but most people would consider it unethical. Similarly, in combat, many actions may be legally justifiable; however, that does not mean they should be taken. Consider the Marine puppy incident: Legally, the act of killing a puppy carries little punishment; however, it is extremely distasteful to most people, and therefore, unethical.

Ethics are an Operational Imperative: Ethical behavior should be encouraged, because it is right. In addition, acting in an ethical manner is operationally smart. The ethical aspect of any battlefield decision can easily take on a strategic implication. Thus, behaving ethically is a "force multiplier," helping warfighters and law enforcement personnel "win the hearts and minds" of the civilians with whom they interact.

How to Maintain Ethics: The tenets of ethical decision-making must be part of all facets of military life and training—in and out of combat. Junior leaders are the key to infusing ethical lifestyles in their teams and units. Below are some suggestions for junior leaders (compiled from the 2010 symposium on *Ethical Decision Making and Behavior in High Performing Teams: Sustaining Future Success*):

- *Discuss Strategic Consequences:* Give guidance and training on the possible 2nd and 3rd order (strategic) effects of decisions. For instance, there may be occasions where strict adherence to ROE (at the tactical level) could still result in an ethical dilemma (at the strategic level).
- *Constantly Emphasize Ethics:* Ethics and ethics training need to be as important as technical, tactical, and physical proficiency.
- *Enforce Standards:* Have the discipline to reward ethical behavior and to punish/penalize bad behavior. These actions need to be seen and projected to all.
- *Clarify "Loyalty" and "Tolerance:"* Loyalty needs to be seen as a hierarchy—to country, to service, to fellow warriors, and then to self. Do not tolerate, by virtue of misplaced loyalty (e.g., to peers), unethical behavior. Do not allow your teammates to act unethically.
- *Instill a Warrior Ethos:* Engender a sense of pride in the trainees' military or law enforcement profession. Communicate the attitude that "we hold ourselves to a high professional standard."
- *Focus on Success:* While there may be ethical lapses in behavior (humans are imperfect), do not dwell on past mistakes. Instead, focus on fixing situations, where needed.

The Abu Ghraib torture and prisoner abuse scandal involved unspeakable illegal and immoral actions taken by US Soldiers. The scandal also set back international relations and progress in Iraq. Archbishop Giovanni Lajolo, foreign minister of the Vatican, is quoted, saying of the incident, "The torture? A more serious blow to the United States than September 11, 2001 attacks. Except that the blow was not inflicted by terrorists but by Americans against themselves."

Discussion Questions:

1. General James N. Mattis has said that "wars among the people will demand ethically sturdy troops." What does this mean? Do you agree or disagree; why?*
2. General Mattis also referred to warfare in the Information Age as a "Battle of Narratives." What does this mean? Why is legal, moral, ethical decision-making so critical in a Battle of Narratives?*
3. It has been said that "an accumulation of tactical victories does not equate to strategic victory." What does this mean, and how does it apply to this discussion on ethical decision-making?

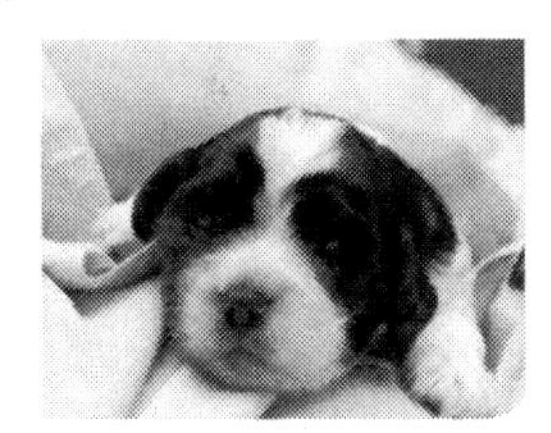

The video of two Marines throwing a puppy off of a cliff raised huge international outcry. Even though this act was not technically illegal, it hurt the Marine Corps' image. The Marines shown in the video were formally punished, and at least one (who is seen throwing the puppy in the video) was discharged.

*Comments made by General James N. Mattis, United States Marine Corps, Former Commander of United States Joint Forces Command, in Washington D.C. on 18 Feb 2010 in a speech, entitled "Developing and retaining the officers we need for the 21st Century."

STRENGTH

Unit 2: Dynamic Decision-Making

2

Under high periods of stress, a person's brainpower narrows. His/her attention shrinks, ability to access knowledge and memories is limited, and capacity to think critically and creatively becomes almost nonexistent. Stress makes making decisions—especially decisions that must be made under the ambiguous and time-sensitive conditions of irregular warfare—extremely difficult. This unit will help trainees understand the dynamic decision-making process. The knowledge and skills gained from this unit can be considered a sort of cognitive toolset that will help warfighters and law enforcement agents "up-armor" their brains for the battlefield and other stressful environments.

Suggested Prerequisites:

- CODIAC Unit 1

Terminal Learning Objective:

- This unit introduces principles related to two of the key training objectives: proactive analysis of dynamic situations and employment of cognitive discipline. At the conclusion of this unit, trainees will be able to explain the cognitive processes associated with the OODA-Loop in more detail, including how these processes are impaired by stress and how decision-making limitations can be mitigated, such as by employing the five combat multipliers.

Enabling Learning Objectives:

- Describe, in moderate detail, the processes that occur at each stage of the OODA-Loop
- Describe some of the limitations that may affect each stage of the OODA Loop
- Describe at least one way to overcome the limitations at each stage of the OODA-Loop
- Describe mental file-folders and how they support dynamic decision-making
- Explain, in one's own words, the concept of Baseline + Anomaly = Decision
- Identify the five combat multipliers
- Explain, in one's own words, how to use the combat multipliers in one's own job

Estimated Time Allotted for Instruction: About 8 hours

Speciality Support Requirements:

- No special facilities required

2001: Situational Awareness

Introductory
Lecture and discussion
About 90 minutes

Terminal Objective: This module includes more detailed information on the observe phase of the OODA-Loop. Upon completion of this module, trainees will be able to describe, in general terms, the process of observation. They will be able to list the five Cooper Color Code, discuss the importance of staying "in the yellow," and articulate ways to help themselves and their teammates maintain high situational awareness.

Module Vocabulary:

- Cooper's Color Code
- Divided Attention
- Focused Attention
- Selective Attention
- Situational Awareness

OBSERVE
ORIENT
DECIDE
ACT

Instructor Activities:

1. Briefly define and discuss attention and the spotlight metaphor
2. Define and discuss *situational awareness*
3. Discuss *Cooper's Color Code*
4. Discuss the importance of team situational awareness

Synopsis of Module Topics:

Introduction: The first stage of the OODA-Loop is observation; in this phase, environmental stimuli are perceived.

Attention: Attention is the process of intentionally concentrating on specific components of the environment while ignoring others. *Divided attention* occurs when two or more things (objects, tasks, etc.) compete for attention, for example, during multi-tasking. *Selective attention* or *focused attention* occurs when distracting or competing stimuli are ignored, in order to maintain attention on a specific object or task.

Situational Awareness: *Situational Awareness* (SA) is an individual's overall understanding of the operational environment, including the time and location of key components, comprehension of their meaning, and a projection of their status in the near future. "Components" may include anything, from friendly and enemy forces and their intent, to the physical terrain features and unit mission status. In other words, SA is internal understanding and integration of the perceived stimuli. SA occurs in one's mind. It is not a display or the common operational picture; it is the interpretation of displays or the actual observation of a situation.

Levels of SA: Commonly, SA is conceptualized at three levels:

- *Perception (Level 1 SA):* Level 1 SA is most basic. It involves observation, cue detection, and simple recognition of situational elements (objects, events, people, systems, environmental factors) and their current states (locations, conditions, modes, actions).
- *Comprehension (Level 2 SA):* Level 2 SA involves the interpretation and evaluation of those stimuli observed at Level 1. This level of SA involves development of a comprehensive picture of the battlespace or operational environment.
- *Projection (Level 3 SA):* Level 3 SA involves anticipation of the future environment, in other words, what will happen next? Level 3 SA is achieved through knowledge of the status and dynamics of the situation (gathered through Levels 1 and 2 SA), as well as how unfolding actions or events will affect the operational environment.

Cooper's Color Code: Developed by Lieutenant Colonel Jeff Cooper, USMC, this system describes levels of awareness. The following Color Code was adapted by the Marine Corps:

- *White:* Unaware and oblivious, personnel in the white state do not notice impending danger; they are unprepared and unready (not attending).
- *Yellow:* A relaxed state of general alertness, personnel in the yellow state are aware of their surroundings; in an operational environment personnel should always maintain a yellow state unless events require a higher degree of attention. (Yellow is "flood-light attention.")
- *Orange:* A heightened state of alertness, in which a specific object or individual is the focus of attention ("spotlight attention").
- *Red:* Ready to fight, personnel in the red condition are mentally prepared for confrontation (like a "laser beam of attention").
- *Black:* Following a catastrophic breakdown of mental and physical performance, personnel can become overloaded and may stop thinking. Leaders must identify personnel in condition Black and take immediate proactive measures to prevent long-term adverse effects. (Personnel in Black have no situational awareness; they offer no tactical value to their units.)

Team SA: While each person must maintain high individual SA, the SA of a collective small unit is equally important. Team SA relies upon team members sharing key information, including their higher level assessments and projections, as well as updates on their own status and capabilities. Team SA can be improved by maintaining constant, effective communication within the team. Communication should be closed-loop, concise, clear, and timely, and it should make use of appropriate jargon. Personnel's updates must also contain information that clearly distinguishes between their observations ("I see a tall person in a berqa") and their interpretations ("I think that may be a man in the berqa").

Discussion Questions:

1. In the Juba sniper video, what Color Code state were most of the Soldiers and Marines displaying? How can you tell?
2. What are some strategies you could use to help your fellow personnel maintain a Yellow status? What could you say to help one of your teammates "snap out of" a White status?
3. What is the danger of being in the Black? What should you do if one of your teammates enters into the Black?

"THE MOST IMPORTANT MEANS OF SURVIVING a lethal confrontation is neither the weapon nor the martial skills: It is the combat mindset."

—*Jeff Cooper*

KEY SKILL

Recognizing when one's situational awareness is low and how to mitigate the condition

Instructors should emphasize the importance of situational awareness, techniques for identifying when one's own or a teammate's awareness is too low, and how to overcome the tendency to "fall into" low levels of awareness over time.

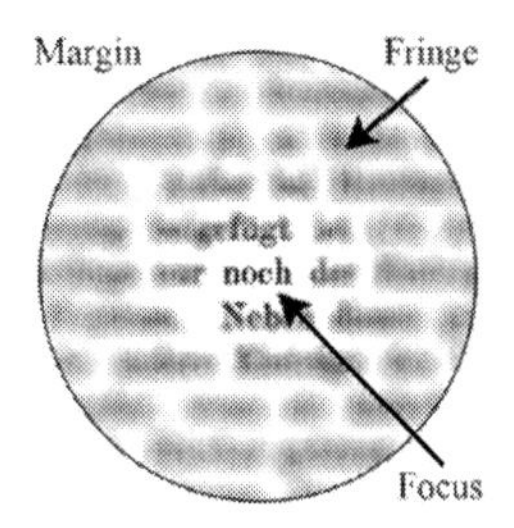

Attention can be conceptualized as a "spotlight," where there is a focused central component, an unfocused fringe, and an ignored margin (LaBerge, 1983)

2002: Sensemaking

Introductory
Lecture and discussion
About 60 minutes

Terminal Objective: This module includes more detailed information on the orient phase of the OODA-Loop. Upon completion of this module, trainees will be able to clearly describe the importance of sensemaking as well as several techniques that they can use to aid individual and team sensemaking.

Module Vocabulary:

- Mental Simulation
- Precipitating Event
- Sensemaking

Resource DVD:

- 2002.1 – *Sensemaking*, PDF

OBSERVE
ORIENT
DECIDE
ACT

Instructor Activities:

1. Define and describe *sensemaking* and its challenges
2. Describe the process of team sensemaking
3. Discuss the usefulness of narrative for sensemaking
4. Discuss the process of mental simulation

Synopsis of Module Topics:

Introduction: The orientation phase of the OODA-Loop is concerned with *sensemaking*, which is "a motivated, continuous effort to understand connections (which can be among people, places, and events) in order to anticipate their trajectories and act effectively" (Klein, Moon, & Hoffman, 2006). In other words, sensemaking is the ongoing process of giving meaning to one's experiences.

Properties of Sensemaking: Karl Weick, one of the foremost sensemaking researchers, suggests that it has seven properties:

- Sensemaking is influenced by one's personal identity
- Sensemaking of the present is always grounded in past experiences
- Sensemaking is constrained by the expected context
- Sensemaking is social; teams create shared meaning and experiences
- Sensemaking is a continuous, ongoing process
- Sensemaking depends upon extracting and interpreting cues
- Sensemaking is driven by assumptions of what is plausible

Sensemaking can be greatly biased by personal (and team) expectations and prior experiences. On the one hand, this speeds the process of sensemaking and is vital for high-tempo environments. On the other hand, personal expectations may limit one's perception of the *real* environmental cues; that is, a person may see what he/she expects to see, rather than what is truly present.

Supporting Team Sensemaking: Three ways to help support team sensemaking in small units include:

- *Shared Responsibility:* Ensure each teammate knows that he/she has the authority—and the responsibility—to bring the team's attention to potential problems or key environmental cues. Make sure the team constantly discusses any potentially relevant cues or anomalies.
- *Attention to Anomalies:* Ensure each person pays special attention to anomalies, those things that fall outside the "normal" pattern.
- *Ambiguity:* Leaders must clarify ambiguous accounts and continuously establish a shared interpretation among the team.

Telling a Story: Stories (or "narratives") tend to follow a predefined structure. This structure aids understanding, because people have a general idea of "how things should go." For instance, most narratives contain a protagonist (hero), an antagonist (villain), and a plot (actions). Creating a narrative of a given operational environment can aid sensemaking. When building an operational narrative, consider these elements:

- *Characters:* Who is involved? What are their relationships?
- *Perspective:* How is one's own perspective biasing interpretation? What might the other characters (such as the enemy) be thinking? What are each of the characters' motivations and goals? How are they feeling at the moment; what is their emotional state?
- *Precipitating Events:* A *precipitating event* is an action or activity that brings about a certain outcome. When identified before a critical event, precipitating events are pre-event indicators. What precipitating events are (or were) observed?
- *Courses of Action:* Finally, what are the potential future courses of action. Step-by-step, visualize each action, reaction, and counteraction, and when possible, explore branches and sequels to the primary plan and identify decision points for critical actions.

Mental Simulation: Operational narratives can be constructed through *mental simulation*, which is the process of imagining how one's predictions about a scene may play-out. When sensemaking, personnel should consciously ask themselves "what if" and consider narrative-based answers. Consider this example: When viewing a city street corner, a police officer sees a suspicious man lingering near a pawn shop entrance. Who might this person be—a criminal, a friend of the pawn shop attendant, a passerby? Next, what might he be thinking and feeling; what does his behavior imply? What pre-event indicators are present; what do these cues suggest? *What if* the man is a criminal, what cues would he give off in that case? Alternatively, *what if* the man is merely a friend of the pawn shop attendant, and he is waiting for his friend's shift to end; if so, what cues would he give off?

Discussion Questions:

1. How might your own experiences and expectations bias your perceptions and sensemaking abilities?
2. Can you think of a real-life example where your sensemaking abilities were impaired because of your expectations? How might you overcome such limitations in the future?

"The attack on the World Trade Center illustrates the sensemaking collapse of the airport security systems, flight crews and passengers, and the first-responder rescue teams. In each case, participants faced unimaginable events, could not recognize the risks, and were unable to act in a cohesive manner to avert danger."

—*Dennis K. Leedom, 2001 Sensemaking report*

KEY SKILL

Generating explanatory storylines that tie individual items of information together

Instructors should emphasize the importance of using mental simulation to help build cohesive "stories" of what the cues in a scene might mean.

KEY SKILL

Imagining alternative courses of action or event outcomes by what-if mental simulations

Instructors should also emphasize the use of mental simulations to help predict future events in a scene. *What if* my team takes this action? *What if* the enemy does this or that? Personnel should constantly look ahead, trying to predict "what will happen next" so that they can act left-of-bang.

2003: Creating Mental "File-Folders"

Introductory
Lecture and discussion
About 90 minutes

Terminal Objective: This module continues to provide detail on the orient phase of the OODA-Loop. Upon completion of this module, trainees will be able to describe, in general terms, how mental "file-folders" facilitate pattern recognition and support sensemaking.

Module Vocabulary:

- Context and Relevance
- Mental File-Folder
- Prototype
- Prototypical Matching
- Schema
- Tactical Shortcut
- Template
- Template Matching

Resource DVD:

- 2003.1 – File-Folders, Video

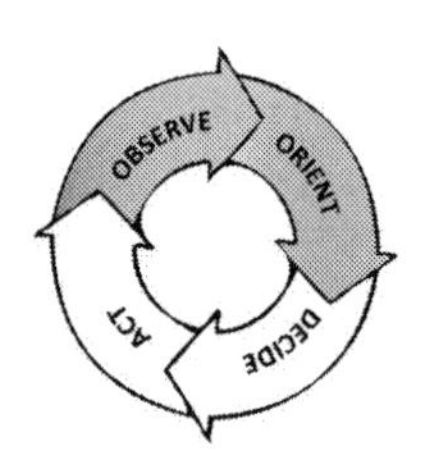

Instructor Activities:

1. Describe *mental file-folders* and explain how they aid sensemaking
2. Describe how file-folders are both tactical shortcuts and biases
3. Review that File-Folders video
4. Compare and contrast *template* and *prototypical matching*
5. Discuss *context and relevance*

Synopsis of Module Topics:

Introduction: As discussed in the previous module, sensemaking is the process of giving meaning to one's experiences. Sensemaking can become more efficient by developing a rich library of mental "file-folders."

Mental File-Folders: A *mental file-folder* (scientifically called *schema*) describes the rules or scripts that people use to structure their knowledge of the world. Individuals develop or refine their mental file-folders each time they interact with the world. Each mental file-folder represents a "normal" experience of a person, place, or thing. For instance, each person has a mental file-folder, or a general sense, for how ordering food at a restaurant works or what to expect when greeting someone new for the first time. Experts have extensive mental file-folders as a result of their numerous experiences. This allows them to better predict what will occur in a given situation and know how to respond most favorably.

Tactical Shortcuts vs. Biases: As discussed in the previous module, sensemaking is influenced by each person's preconceived expectations and prior experiences. Mental file-folders create these expectations, for better (accelerating decision-making) and worse (creating biased perceptions). The type of training that personnel receive directly influences how they create their mental file-folders, and therefore how easily file-folder information can be retrieved. Negative training, singularly focused training, or incomplete training can create corrupt file-folders that lead to negative biases. However, with proper focused training experiences, file-folders can become accurate scripts for individuals' actions. In other words, they can become *tactical shortcuts* that drive decisions.

Template Matching: A *template* is a design or pattern that guides the design or construction of identical items. In other words, a template is an exact specification. Similarly, in regard to decision making, *template matching* occurs when a person looks for an exact match. For instance, if a Solider searches for a particular Person of Interest, then that Solider is looking for an exact, or template, match. When people look for exact matches, however, they create a bias against finding anything else, which can result in "change blindness" (discussed in more detail in Unit 3).

Prototypical Matching: A *prototype* is an original form or instance of something, and it serves as a typical example for items of the same category. In regard to decision-making, a *prototypical match* is a "close enough" match. For instance, if a police are looking for members of a certain criminal gang, officers may have general characteristics for which they are looking, such as particular tattoos or clothing styles, but they might not necessarily be looking for specific individuals. When possible, prototypical matching should be used; in other words, personnel should try to look for general characteristics in individuals or settings, rather than looking for limited predetermined characteristics. This can help personnel maintain greater situational awareness, accelerate sensemaking, and help guard against change blindness.

Context and Relevance: *Context* is the background, setting, or situation surrounding an event or occurrence, and *relevance* describes whether something is significant or meaningful to a given situation. Context and relevance are important for sensemaking; by keeping context and relevance in mind, personnel will be better able to look for anomalies, or things that are out of place. For instance, in some places in Iraq, warfighters found blue barrels filled with thousands of washing machine timers. In an American industrial area, a barrel full of washing machine parts might not be relevant. However, in Iraq, a context in which practically no one owns a washing machine, barrels full of washing machine timers is an anomaly. Even a house with one or two washing machine timers is a deviation from the norm. Washing machine timers' relevance is that they are often used to delay ignition in IEDs; hence, individuals in possession of timers are likely involved with the insurgency.

KEY SKILL

Inducing a pattern from a few individual cues

Instructors should emphasize the importance of intuiting a meaningful pattern from observing just a few relevant cues. Personnel can achieve this, in part, by maintaining awareness of a situation's context, considering the relevance of specific cues, looking for prototypical (rather than template) matches, and developing an extensive library of operationally relevant mental file-folders.

Discussion Questions:

1. How can a trainer ensure that trainees have enough appropriate experiences so that they develop accurate mental file-folders? How might trainers help convey the key "context and relevance" of a training experience? What techniques could a trainer use to compare-and-contrast training experiences?
2. Why are mental file-folders important for personnel to develop? How can they aid decision-making in irregular or ambiguous environments?
3. Different people have different backgrounds. Some personnel grow-up in cities, others in rural areas, and so on. How does someone's background affect his/her file-folder "rolodex"? Is it useful to work with a personnel from different backgrounds? How might different personnel's prior experiences be best employed during operations?

2004: Dynamic Decision-Making

Introductory
Lecture and discussion
About 90 minutes

Terminal Objective: This module includes more detailed information on the decide phase of the OODA-Loop. It lays out, in general terms, the challenges associated with making decisions in irregular and ambiguous operational environments. Upon completion of this module, trainees will be able to describe the cognitive limitations that occur under stressful and ambiguous conditions, and they will be able to list a few strategies for overcoming those challenges.

Module Vocabulary:

- Automaticity
- Cerebral Cortex
- Fight, Flight, or Freeze
- Limbic System
- Memory–Emotion Link

Resource DVD:

- 2004.1 – The Brain, Video
- 2004.2 – *Naturalistic Decision Making*, PDF

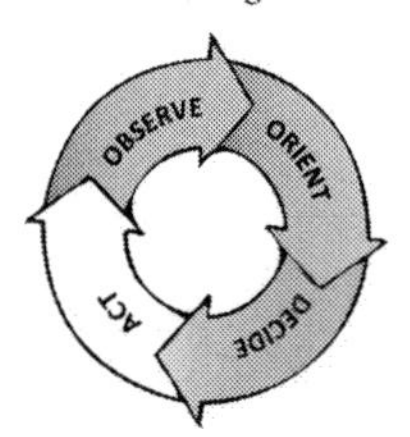

Instructor Activities:

1. Define dynamic decision-making
2. Describe the challenges associated with dynamic decision-making
3. Describe, in general terms, how the brain functions under stress
4. Describe the importance of *automaticity*
5. Discuss the *memory–emotion link*

Synopsis of Module Topics:

Introduction: Today's operational environments—whether in a military theater, on the US border, or in an inner city—tend to be ambiguous, dangerous, and dynamic. Such contexts place personnel under high degrees of stress. Thus, the nature of today's conflicts, along with their added stress, challenge decision-making.

Dynamic Decision-Making: Dynamic decision-making takes place when a series of actions must be executed over time, the actions are interdependent, the outcomes of actions are not immediately evident, and the environment is changing both spontaneously and in response to actions.

- *Distributed Over Time:* Actions must be executed over time, so there are many "moving parts" to orchestrate and observe. This challenge can be mitigated, in part, by maintaining high situational awareness among individuals, within teams, between teams, and between teams and the tactical operations center.
- *Interdependency:* Actions are interdependent, which means one person or team's successes or failures affects everyone else. However, the interdependency of dynamic situations can be used to positive effect; personnel in cohesive teams can support one another, helping each other maintain situational awareness and providing back-up support to one another.
- *Delayed Feedback:* The outcomes of actions are not immediately apparent. For instance, when a squad hands out aid in Afghanistan, whether this action positively impacted the attitudes of the community may never be truly known. Delayed feedback makes decision-making extremely challenging. Personnel must rely upon mental simulations and their operations centers' intelligence pictures.
- *Ambiguous, Evolving Environment:* The outcomes of a specific operation may not have the greatest effects in a particular environment. In the above example, handing out aid might positively affect the tribe, but an unrelated event may instill negative attitudes. As stated above, mental simulations and collaborative sensemaking must be used to overcome this challenge.

Decision-Making Under Stress: Dynamic environments also involve time pressure, and for military and law enforcement personnel, they often involve hostile conditions. Consequently, dynamic environments are extremely stressful. When individuals are faced with extreme stress or a violent crisis, the brain's *limbic system* takes control from the *cerebral cortex*. The limbic system is an "older" part of the brain (in terms of evolution) and is involved in instinctive behavior and emotions. The cerebral cortex, in contrast, supports human cognitive abilities such as logical thinking, reasoning, and analysis. Therefore, when the limbic system takes over during stress, instinct (not careful thought) drives an individual's actions. When faced with high stress situations, a person will either:

- *Fight:* Respond with immediate, aggressive response
- *Flight:* Flee or attempt to avoid the perceived threat
- *Freeze:* Hesitate and freeze, often in a state of initial denial

For military and law enforcement personnel, it is particularly important to train their minds to react appropriately under stressful conditions. Physical—and mental—conditioning will assist personnel in overcoming the negative effects of stress. This can be achieved through training in realistic scenarios, exercising decision-making process through tactical decision-making games, and training under stressful, demanding situations.

Automaticity: *Automaticity* means learning of a task to a point that it becomes essentially attention-free. This is why military personnel practice gun-drills repetitively, so that they do not have to think about these actions once they enter the battlespace. One goal of CODIAC training is to make enhanced observation and decision-making more automatic.

Memory–Emotion Link: Emotional responses can create highly durable memories. If individuals make strong enough emotion–memory links during training, they will instinctively refer to that training during periods of high stress. Whenever individuals have an emotional tie to an event, object, person, or information, it will be more easily remembered because it involves the more "instinctive" or "emotional" parts of the brain (limbic system) as well as the more "rational" parts of the brain (cerebral cortex). However, take care to always train the rational *and* instinctive parts of the brain, because the strength of emotion can obscure memories, making them less accurate.

Discussion Question:

1. How might you determine if one of your teammates has become overloaded by acute stress? How could you help him/her regain composure? Name three techniques you could use in a real-world setting.

KEY SKILL

Making effective decisions in spite of high stress conditions

Instructors should emphasize the importance of making good decisions, despite stress. Personnel must be prepared to experience stress, recognize its affects, and overcome the natural physiological limitations that stress imposes.

KEY SKILL

Recognizing when stress is affecting other team members' actions and helping them refocus their attention

Similarly, instructors should encourage trainees to look for signs of stress in their teammates. Personnel should be prepared to help their teammates recognize and overcome the immediate affects of acute stress.

ACUTE STRESS

Acute stress occurs when a situation involves high physiological arousal, requires individuals to make multiple decisions rapidly, has only incomplete information available (high uncertainty), and has outcomes that involve extreme consequences, such as life or death (Salas, Driskell, & Hughes, 1996).

2005: Baseline + Anomaly = Decision

Introductory
Lecture and discussion
About 60 minutes

Terminal Objective: This module introduces trainees to the concepts of *baseline* and *anomaly*, as well as how, together, they require a decision. Upon completion of this module, trainees will be able to describe a baseline, an anomaly, and how anomalies to the baseline require a decision.

Module Vocabulary:

- Anomaly
- Baseline

Resource DVD:

- 2005.1 – Baseline, Anomaly, Decision, Video
- 2005.2 – Decisions, Video

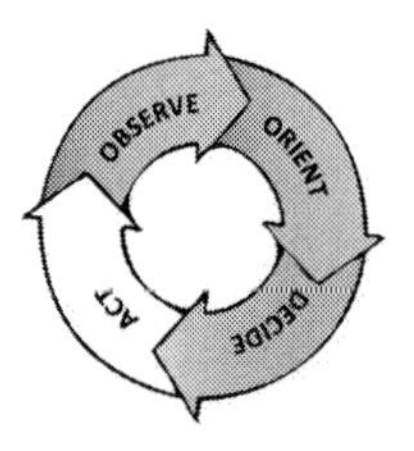

Instructor Activities:

1. Describe the concept of a *baseline*
2. Describe the concept of an *anomaly*
3. Review the videos provided on the Resource DVD
4. Describe why a decision is required when there is an anomaly

Synopsis of Module Topics:

Introduction: Successfully navigating, surviving, and dominating the tactical environment requires personnel to be hyper-aware of the natural state of the area of operation, so that they can swiftly detect any subtle cues when they occur. These cues are the indicators that enable a trained warfighter to detect and react to danger left-of-bang.

Baseline: A *baseline* is the "normal" status of a thing. Everyone and every place has a baseline, although, baselines are fluid, changing over time as conditions evolve. For example, a certain village may have a large field in which children play. If personnel watch this village over time and every afternoon the children play in the field, then those children and their activities become part of the baseline for that village. They become part of personnel's mental file-folders for what is normal.

Anomaly: An *anomaly* occurs when something above or below the expected baseline happens. An anomaly above the baseline represents something added to the environment, such as the presence of a new vehicle in a neighborhood. An anomaly below the baseline represents the absence of something, such as the lack of wildlife noises in the woods.

When a warfighter observes an anomaly, he/she must analyze it to determine whether it is operationally relevant. In the above example, the baseline is that children play in the field. If, one day, a squad leader observes that no children are in the field, he/she must determine the significance of that observation, considering the context and relevance of the situation. For instance, lack of children could mean something mundane; maybe the children are attending a party in another part of the village. Or it could indicate danger; perhaps the local population knows a hostile force is in area, and the children have been kept inside to avoid danger.

Action: All anomalies should be investigated; in other words, never "write off" a single anomaly. Once personnel identify an anomaly, they should heighten their awareness, looking for more clues. If three anomalies are observed, then personnel must act. Similarly, if one major anomaly is identified, action must be taken. Remember, as discussed in the previous module, cues must be considered in conjunction. What pattern do multiple cues imply? What narrative do the combined cues imply?

Combat Profile: When an existing file-folder matches a known or suspected nonstandard observation, then the match becomes a predictive match or a *combat profile*. A combat profile is a model for comparison against future or suspected future events.

Using Baseline + Anomaly = Decision: Consider this example. A Marine develops a mental file-folder during training that connects washing machine timers with IEDs. Later, in Iraq, that Marine sees a washing machine timer in a house he is searching. He has identified an anomaly, but a single washing machine timer is not enough evidence to act. So, the Marine begins looking for additional anomalies. He notices that the wall around the courtyard is marked with some subtle graffiti; although he cannot read the graffiti, he realizes this is a second anomaly. Finally, upon closer inspection, the Marine finds a dozen batteries, of various sizes, in a drawer in the home. Alone, each of these cues means little, but the three anomalies, together, form a picture. They create a story that suggests the resident of the home is involved in IED creation. The Marine now has a combat profile in mind; he decides to "capture" the homeowner and ask a tactical questioner to interview the man.

Homemade Explosives: Improvised Explosive Device (IEDs) and Homemade Explosives (HME) are responsible for approximately 40% of all causalities in the current military conflicts according to the Iraq Coalition Casualty Count. IEDs are also being proliferated along the northern Mexican border. Personnel must learn the cues that suggest a suspect is involved with HME or IED creation. Discuss these threats whenever possible and use HME and IEDs in class discussions and examples.

Discussion Question:

1. How does one establish a baseline for an area of operations? How does one continually update the baseline?

Left: Explosive devices being destroyed by specialists at Sather Air Base, Iraq. (U.S. Air Force photo)

"WE SAY IN COMBAT 'complacency will get you killed.' What does complacency mean? Complacency means that you relax too soon—that you're not paying attention to the environmental cues."

—Greg Williams

KEY SKILL

Establishing a baseline to extract normalcy

Instructors should emphasize the importance of establishing a baseline. Discuss the concept of a baseline with enough detail that all trainees fully understand this critical concept, and discuss ways to create, maintain, and update mental representations of baselines.

KEY SKILL

Looking for anomalies outside of the baseline

Again, instructors should emphasize the importance of identifying anomalies that fall outside of the baseline. Recognizing and interpreting anomalies from the baseline is *the* core CODIAC skill.

2006: Combat Multipliers

Introductory
Lecture and discussion
About 60 minutes

Terminal Objective: This module introduces trainees to the concept of combat multipliers. Upon completion of this module, trainees will be able to discuss and describe the various combat multipliers available to the tactical team, their advantages, and how a team can employ them to best effect.

Module Vocabulary:

- Combat Multiplier
- Good Shepherd
- Guardian Angel
- Interlocking Lines
- Overwatch
- Tactical Cunning
- Tactical Patience

Resource DVD:

- 2006.1 – Multipliers, Video

Instructor Activities:

1. Review the Multipliers video
2. Define and discuss *tactical cunning*
3. Define and discuss *tactical patience*
4. Define and discuss *interlocking lines*
5. Define and discuss *guardian angel*
6. Define and discuss *good shepherd*
7. Discuss how and why these combat multipliers help personnel

Synopsis of Module Topics:

Combat Multipliers: *Combat multipliers* increase a unit's combat effectiveness while the actual force ratios remain constant. Examples include tactics, leadership, munitions, optics, training, and morale. CODIAC-related combat multipliers include tactical cunning; tactical patience; interlocking observation, reporting and fires; guardian angel; and the good shepherd concept.

Tactical Cunning: *Tactical cunning* is the process of out-thinking the enemy. Possessing tactical cunning requires that personnel think like the enemy and then use crafty strategies to surprise an adaptive enemy, so that they cannot anticipate the actions of friendly forces. Ways to improve tactical cunning include:

- Consider the Enemy's Most Probable Course of Action (EMPCOA) and the Enemy's Most Dangerous Course of Action (EMDCOA).
- Think about what the enemy expects might be *your* most probable or most dangerous course of action.
  - — Use deception to mislead an enemy, manipulating cues to induce him/her to react in a manner prejudicial to his/her interests.
  - — Use ruses to deceive an adversary, such as exposing false information to the adversary's intelligence system.
  - — Use feints to deceive an adversary in offensive operations, disguising the location and/or time of the actual main offensive.

Tactical Patience: *Tactical patience* is the manipulation of the operational tempo, so as to act only once the situation becomes most advantageous. For instance, while conducting an ambush, Soldiers may allow an enemy advanced guard to pass the ambush site in order to destroy the main body in the kill zone. Additionally, sometimes delays at the tactical level can lead to greater successes at the strategic level; for example, instead of apprehending an insurgent immediately, warfighters may follow him, identifying the people with whom he works before capturing the man.

Interlocking Lines: By *interlocking lines* of fires, observation, and reporting, personnel can close the seams and gaps of a battlespace. Most likely, personnel are already familiar with the concept of interlocking fields of fire. Interlocking fields of observation are similar to this concept, and interlocking lines of reporting shares a conceptual similarity.

- *Interlocking Fields of Fire:* Position the arcs of weapons' fire so that friendly teams are always fully covered.
- *Interlocking Observation:* Similarly, position personnel and control their optics use, so that teams can collectively create a complete observational picture of a particular person or site.
- *Interlocking Reporting:* Finally, small unit leaders must ensure that communications are transmitted between different units (higher, adjacent, and subordinate) to create an accurate common tactical picture of the battlespace. It is critical to share information laterally to other units, in addition to one's higher and subordinates.

Guardian Angel: *Guardian angels* are alert personnel, placed in covert positions, to protect their units and all the while remaining unseen by the enemy. Each unit establishes patterns by its very existence, and the enemy responds to those patterns. Eventually all units will be spotted and their patterns discerned by the enemy. Therefore, tactical teams must always have at least one guardian angel who the enemy cannot find, that is, at one person in a position of ambush, providing *overwatch* for the rest of the unit. For warfighters, in particular, security is the first priority; consequently, guardian angel placement must also become a priority. It should become a habit of action and be instilled via a memory–emotion link.

Good Shepherd: Being a *good shepherd* is the process of building trust within a local population, winning their "hearts and minds." "Hearts" means persuading people that their best interests are served by the success of friendly forces, and "minds" means convincing them that friendly forces can protect them. Note that neither concept has to do with whether people like the coalition forces or law enforcement personnel. Instead, they are focused on building trust, security, and stability.

Discussion Questions:

1. Consider the area of operations for your unit. What combat multipliers can be employed to increase the combat strength of your unit?
2. Think back to the Juba sniper video. Could these combat multipliers have helped those Soldiers and Marines be more secure? What if those Soldiers and Marines had employed guardian angels?

KEY SKILL

Using tactical patience to avoid committing too soon or going to kinetics unnecessarily

Instructors should emphasize the importance of tactical patience. Describe the concept thoroughly to trainees, and then discuss scenarios in which tactical patience was (or could have been) used to support mission success.

KEY SKILL

Using geometry of fires to create an interlocking network of optics, intel, and communications

Similarly, instructors should emphasize the importance of interlocking lines. As above, describe this concept thoroughly, and then discuss scenarios in which interlocking lines of fire, observation, and reporting were (or could have been) used to support success.

This is your true main effort: everything else is secondary. Actions that help build trusted networks serve your cause. Actions—even killing high-profile targets—that undermine trust or disrupt your networks help the enemy.

—*Lt Col David Kilcullen, Australian Army*

Unit 3: Enhanced Observation

3

Observation is the gathering and processing of information obtained through the senses. *Enhanced observation* involves a deeper understanding how human sensory systems work, the limitations of these systems, and how to mitigate one's own perceptual limitations while exploiting the enemy's observational weaknesses. In this unit, trainees will learn how to become more effective observers by gaining a more detailed understanding of human observation, observational equipment, and techniques for defeating enemy observation.

Suggested Prerequisites:

- CODIAC Units 1–2
- Trainees should already have basic familiarity with their organic optical devices (i.e., "knob-ology")

Terminal Learning Objective:

- This unit introduces principles related to the major training objective of using enhanced observation techniques. At the conclusion of this unit, trainees will be able to briefly describe the physiological and cognitive processes involved with observation and perception. They will be able to succinctly list the limitations of the eye and mind, how these limitations contribute to perceptual errors, and how to overcome these limitations with enhanced observation techniques.

Enabling Learning Objectives:

- Explain, in general terms, how the eye senses or observes the world
- Explain, in general terms, how the brain perceives the world
- List the factors that influence how easily something is seen
- Describe the natural limits of observation and perception and how they can be minimized
- Demonstrate the use of enhanced observation techniques with one's own optical devices

Estimated Time Allotted for Instruction: About 10 hours

Speciality Support Requirements:

- All trainees should have access to their organic optical devices, which may include (but should not be limited to) *day optics*, such as Rifle Combat Optics (RCOs) and binoculars, *night optics*, such as a Night Optical Devices (NODs), and *thermals*.
- The two demonstrations in this unit require several role-players, a range, and support equipment. Details are provided in Module 3006.

3001: Principles of Observation

Introductory
Lecture and discussion
About 60 minutes

Terminal Objective: This module introduces trainees to the physiological (i.e., eye) and cognitive (i.e., brain) processes involved with sensation. Upon completion of this module, trainees will be able to describe, in general terms, how the human eye observes the world. They will be able to list some natural limitations of visual perception, as well as identify visual features to which the eye is easily drawn.

Module Vocabulary:

- Cone Cells
- Cornea
- Fovea
- Perceptual Fill
- Retina
- Rod Cells
- Sensory Systems

Resource DVD:

- 3001.1 – The Eye, Video

Instructor Activities:

1. Describe how the human eye senses the world
2. Discuss the limitations of human sight
3. Review the Eye video
4. Discuss factors that influence how easily something is seen
5. Lead an interactive discussion on how the physical aptitudes and limitations of the human eye may affect operations

Synopsis of Module Topics:

Introduction: Observation begins by gathering and processing information through the senses (sight, hearing, smell, touch, taste). While all senses are important, personnel tend to rely most upon the sense of sight.

Structure of the Human Eye: It is important for personnel to understand, in general terms, how the eye works so that they can make optimal observations and overcome natural limitations of the eye.

- *Rods vs. Cones:* The eye contains two types of light-sensing cells: *rods cells* and *cone cells*. Rod cells see only black-and-white. They activate in low light and support peripheral vision. Consequently, both night vision and peripheral vision lack color. Cone cells see in color, but they require a lot of light to activate. (Ice cream cones come in a variety of colors, and the cone cells in one's eye see in color. This mnemonic may help trainees remember the distinction between rods and cones.)
- *Central Vision:* A tiny spot (scientifically called the *fovea*) located in the middle of each eye is responsible for sharp, central vision. This tiny spot is where all visual detail is detected; in contrast, peripheral vision is much less focused. Central vision is about the size of a quarter at one arm's length.
- *Blind Spot:* Each eye contains a blind spot where nerves enter the eye. This means that there is an area of the visual field (the area of space that can be seen at one time) that is not seen. Yet, this does not appear as a hole in one's vision. The brain does not like incompleteness, so it fills in what it "thinks" it sees. This is called *perceptual fill*.

Bias Towards Motion: Human eyes are predisposed to notice moving, rather than stationary, objects. The perceptual system is particularly sensitive to biological motion (the movement of living things). An individual can identify gender, activity, and even the identity of a person through movement cues alone. For example, a teammate may be able to identify another member of his/her squad from a long distance, just watching the way that person moves.

Central vs. Peripheral Vision: Humans have only a small area of central (foveal) vision. The rest of the visual field falls under peripheral vision. Peripheral vision will usually not sense an object that remains still but will be drawn to anything moving. Paying attention to the periphery helps personnel maintain greater overall awareness, but this places high demands on the attentional system and cannot be maintained for a long time.

Daytime vs. Nighttime Vision: Daytime vision (called *photopic vision*) becomes active under well-lit conditions and relies on the eye's cone cells. This system is most sensitive to wavelengths of light that correspond to the yellow/green spectrum. Many fire trucks are now being painted these colors because they are the most visible colors under human photopic vision. Nighttime vision (or *scotopic vision*) activates under darker conditions, and it relies on the eye's rod cells. The rods are only sensitive to short (blue) wavelength light, and because they are not sensitive to other wavelengths, the scotopic system does not have color vision.

Dark Adaptation: Total dark adaptation (being able to "see" in the dark) takes approximately 30 minutes to achieve. For the first 10 minutes in darkness, the cones cells maintain control of the visual system; after that time the rods take over. Once the rods activate, an individual becomes more sensitive to light and can begin to see in lower-light conditions. For example, when a person first walks into a dark movie theater he/she cannot see very well, but after ten minutes of sitting in the dark room it becomes easier to see clearly.

Color Vision: Colors appear to change as light conditions change, because of the differential use of rod and cone cells. Reds, oranges, and yellows appear relatively light in bright illumination, whereas blues appear relatively light in dim illumination. Technically, this phenomenon is called the *Purkinje Shift*. In practice, it means that colors can appear to change and care must be taken to correctly identify items by color alone.

Twilight Vision: Twilight (or *mesopic*) vision is used under medium light conditions (e.g., at twilight), and it involves both rod and cone cells. The combination of photopic and scotopic systems can cause visual inaccuracies, making personnel most susceptible to attack during this time.

Discussion Question:

1. Consider the limitations of the human eye. How might these limitations affect reconnaissance and surveillance activities? How might you overcome these obstacles?
2. When is the most vulnerable time of day, as far as vision is concerned?

Why Things are Seen:

Certain properties will readily draw the eye. These include:

- *Shape:* Familiar things have recognizable shapes, including the outline of individuals, man-made objects, or other geometric shapes that do not occur in nature
- *Shine:* Reflected light attracts the eye
- *Color:* The greater the contrasting color, the more visible the object becomes
- *Shadow:* In sunlight, an object or person will cast a shadow, which may be more revealing than the object or person directly
- *Silhouette:* Contrasting colors and shapes that break the skyline; look for silhouettes at the contrast between positive and negative space
- *Surface:* If an object has a surface that contrasts with its surroundings, it becomes conspicuous
- *Spacing:* Nature never places objects in regularly spaced patterns. Only humans use rows and equal spacing
- *Movement:* Even when all other indicators are absent, movement will give a position away.

3002: Human Information Processing

Introductory
Lecture and discussion
About 120 minutes

Terminal Objective: This module offers an overview of human information processing and perception. Upon completion of this module trainees will be able to describe, in general terms, how the brain interprets visual stimuli and what natural limitations may hinder the perceptual process.

Module Vocabulary:

- Bottom-Up Processing
- Chunk
- Cognitive Load
- Long-Term Memory
- Mnemonics
- Perception
- Top-Down Processing
- Working Memory

Instructor Activities:

1. Describe the differences between *observation* and *perception*
2. Describe the information processing model of human cognition
3. Discuss *top-down/bottom-up processing* and *cognitive load*
4. Explain and discuss the use of *mnemonics*

Synopsis of Module Topics:

Introduction: As discussed in Unit 2, sensory systems participate in the observation phase of the OODA-Loop, and then the brain becomes more involved during the orient phase. During the orient phase, the brain begins to perceive the stimuli that the eyes, for instance, have sensed. *Perception* is the process by which sensory information is organized and interpreted to produce a meaningful experience of the world.

Information Processing Model: Human information processing theory helps explain how people receive, store, and use information. Whole branches of science are dedicated to understanding human information processing; only a brief description is provided here:

- *Sensory System:* As mentioned, stimuli are first captured by a sensory system, such as the eyes (sight) or ears (hearing).
- *Sensory Memory:* The observed stimuli then enter sensory memory, where they are remember for several milliseconds. If the brain decides to pay attention to the stimuli, then they enter *working memory.*
- *Working Memory:* Working (or short-term) memory handles the interim processing. Information in working memory is stored for only a few seconds, unless it rehearsed. For instance, if a person has to remember a phone number, he/she might repeat it until he/she can type the digits into to a phone. Working memory is also limited by capacity; under ideal conditions, it can only store about seven plus-or-minus two (7±2) pieces of information at a time. For these reasons, working memory is a substantial "cognitive bottleneck." Information that exceeds this bottleneck is shed, or in other words, forgotten.
- *Long-Term Memory:* If retained by working memory, information enters *long-term memory*. Long-term memory has a theoretically unlimited capacity; however, information stored in long-term memory cannot always be remembered (or "retrieved"). For instance, people may experience the tip-of-the-tongue phenomenon, where they know a fact but cannot access it. In order to support retrieval from long-term memory, new information must be integrated with previous knowledge. This forms a sort of "web" of knowledge.

Top-Down vs. Bottom-Up Processing: Perception driven by the features of stimuli (such as color, motion, shape) is called data-driven or *bottom-up processing*. In bottom-up processing, stimuli are primarily interpreted based upon the data gathered by the five senses, rather than by preexisting expectations. In contrast, schema-driven or *top-down processing* is guided by expectations, such as existing mental file-folders.

- *Top-Down Perception:* When available, sensory memory references existing knowledge in order to determine to which cues to attend. In other words, top-down perception helps determine what information is passed to working memory. Experts with extensive prior knowledge can look at a scene and immediately notice the most important cues.
- *Bottom-Up Processing:* Novices, on the other hand, lack experience. Their perception is instead guided by visual exploration and perceptive salience; that is, their attention is drawn to those features that visually standout, whether or not they represent the most critical cues in the scene.

"THESE THINGS WILL GET US KILLED. When we are in an environment that is target rich and there is a whole bunch of sensory input, we will look—just like a trout—at all those things that are flashing around us, and we'll take our eyes off the prize."

—*Greg Williams*

Cognitive Load: *Cognitive load* describes the load on the information-processing system, especially working memory. Since working memory is limited by size and duration, humans can only processes a certain amount of information at a given time. Experts appear to process more information at a time, because they use top-down processing and mental file-folders, which allow them to monitor only the important information and to "*chunk*" (or cluster) bits of information together.

Mnemonic Devices: *Mnemonics* are mental tricks that aid memory and retention. Usually, mnemonics rely on easy-to-remember information that creates a framework for the important (but easily forgettable) data someone wants to recall later. Personnel can use mnemonics to help remember key details of a scene, specific procedures, or critical information gathered from discussions with villagers.

Discussion Questions:

1. What is the difference between observation and perception?
2. What is the most critical "cognitive bottleneck"? Why should personnel be aware of this bottleneck? What can personnel do to compensate for the limits of working memory?
3. People often use mnemonics, such as creating a memorable phrase from the first letters of items. For instance, individuals may remember the planets by memorizing the phrase "My Very Educated Mother Just Served Us Nachos." What other mnemonics might be useful?

3003: Overcoming Limitations

Introductory
Lecture and discussion
About 60 minutes

Terminal Objective: This module introduces trainees to several natural limitations of the sensory and perceptual systems. Upon completion of this module, trainees will be able to discuss the training and techniques required to overcome these limitations.

Module Vocabulary:

- Binocular Vision
- Change Blindness
- Channel Capacity
- Cognitive Illusion
- Focus Lock
- Monocular Vision
- Sequencing
- Tunnel Vision

Resource DVD:

- 3003.1 – Dan Ariely, Video
- 3003.2 – Overcoming Cognitive Limitations, Video

Instructor Activities:

1. If time permits, review the Dan Ariely video (from TED Talks)
2. Review the Overcoming Cognitive Limitations video
3. Describe the physiological limitations of observation
4. Discuss training and techniques for overcoming these limitations

Synopsis of Module Topics:

Introduction: The last two modules briefly described human sensation and perception. This section expands upon the limitations of these systems and describes how such limitations can lead to observation errors.

Limitations: All humans have natural physiological limitations. Through training, warfighters and law enforcement agents can understand these limitations and learn skills that help overcome them.

Limitations of observation include the following:

- *Fatigue*: Constantly viewing an area, particularly through optical devices, will quickly fatigue the eye. Personnel should frequently switch between optical devices and regular vision. Also, even under ideal conditions, personnel should rotate out of observation duties at least every 20–30 minutes.
- *Monocular Vision:* Most visual perception is a combination of sight from both eyes; this is *binocular vision*. Objects seen with binocular vision are perceived in three dimensions. However, objects seen with only the left or right eye are perceived in two dimensions; this is *monocular vision*. Binocular vision helps humans perceive depth, which is why depth perception becomes hindered when personnel use an optical device with just one eye.
- *Tunnel Vision:* During periods of high stress, people may develop *tunnel vision*. Physiologically, tunnel vision literally means reduced peripheral vision. The phrase is also used metaphorically to imply that individuals are attending to fewer cues and ignoring important tasks.

Limitations of perception include the following:

- *Change Blindness:* Humans may ignore stimuli (even quite obvious stimuli) when their attention is focused elsewhere. This phenomenon is called *change blindness*. To overcome change blindness, it is helpful to memorize or sketch what the observed area looks like, and then refer back to this baseline at later times. By comparing what one is currently viewing with what was memorized/sketched 20 minutes prior, the brain is better able to perceive changes.

- *Focus Lock:* When observation becomes fixated on a specific object this is called *focus lock*. Efficient use of refocusing in visual scans, moving one's sight to include both near and far objects, and constantly shifting one's field of view can help mitigate focus lock.
- *Channel Capacity: Channel capacity* is the maximum data rate that can be maintained over a given channel. Under normal conditions, humans' channel capacity is about seven plus-or-minus two (7±2); however, under stress humans' channel capacity drops to about three. In other words, under challenging conditions, personnel can only pay attention to about three things.
- *Sequencing*: Human brains try to place cues into grouping (even inaccurate groupings). *Sequencing* occurs when our brains create an inaccurate grouping based upon a sequence of observed cues; typically sequencing takes place at the seventh instance of a cue. For instance, if a police officer makes six traffic stops without incident, then his/her brain will relax—it will assume the next traffic stop will play-out like the previous six. This is why so many officers are killed-in-action during their seventh traffic stop. If at all possible, personnel at checkpoints or conducting traffic stops should be rotated every six instances.
- *Adaptation:* People have a tendency to ignore visual information that is continuously present because the brain "tunes it out." This *adaptation* can be dangerous; for instance, if graffiti is always present, personnel may begin to overlook its potential importance, even if a minor change occurs that could hold tactical significance. Personnel must remember to use their observation skills to ensure they do not ignore potentially meaningful changes in the "normal" environment.

THE LONGER WE'RE IN COMBAT, the less we pay attention. ...Terrorists and insurgents live off of that. They know that the longer you're on site, the less you're paying attention.

—*Greg Williams*

KEY SKILL

Shifting field of view to avoid focus lock

Instructors should emphasize the importance of avoiding focus lock. Discuss conditions under which focus lock might occur, and ask trainees if/when they have personally experienced it. Discuss strategies for reducing the likelihood of focus lock.

Overcoming Cognitive Illusions: Many of these limitations can be considered *cognitive illusions*. Cognitive illusions occur when the brain makes (incorrect) unconscious inferences. In general, cognitive illusions can only be overcome through experience and training.

Rely Upon Teammates: In addition to honing individual observation skills, personnel must acknowledge the value of their team. A tactical team can create a common tactical picture, remind each other to avoid natural limitations, and backup each other to minimize the danger of the natural limitations of humans' sensation and perception systems.

Discussion Question:

1. Consider the area and types of mission in which your unit engages. What kinds of observational training will best overcome the limitations you will encounter in that area and for those types of missions?

3004: Observation Techniques

Introductory
Lecture and discussion
About 60 minutes

Terminal Objective: This module introduces trainees to observation techniques. Upon completion of this module, the trainee will be able to describe and demonstrate the two primary types of observation techniques and how to maintain observation.

Module Vocabulary:

- Detailed Search
- Hasty Search
- KOCCOA
- Negative Space
- Overlapping Strip Method
- Positive Space

Instructor Activities:

1. Describe the purpose of observation techniques
2. Describe the *hasty* and *detailed search* techniques
3. Discuss the kinds of "key features" warrant extra attention
4. Describe maintaining observation

Synopsis of Module Topics:

Introduction: When a tactical team enters a target area or battlespace, one of its first priorities should include scanning for hostile forces. The tactical team should assign several observers to cover overlapping fields of observation and conduct searches of their visual fields.

Hasty Search: In a *hasty search*, the observer quickly glances at specific points, terrain features, or other areas that could conceal the enemy. He/she does *not* sweep his/her eyes across the entire terrain, because that is less effective at detecting movement. Instead, the observer should start the search by viewing the area closest to his/her position and then working out. Searches should be conducted by looking from *right-to-left* (because this goes against most people's natural tendency to read from left-to-right). If able, personnel should use binoculars, rather than higher powered optics, during a hasty search. Binoculars give the observer a wider field of view, thereby increasing the efficiency of the search. If a threat is detected, the observer should transition to a higher power optic, if available.

Detailed Search: After completing a hasty search, the observer should initiate a detailed search using the *overlapping strip method.* Normally, the area nearest the observer offers the greatest danger; therefore, the search should begin there. The observer systematically searches the terrain, starting at the right flank and then moving his/her observation towards the left in a 180° arc. Each visual arc should include about 50 meters of depth. After reaching the left flank, the observer searches the next swath nearest to his/her post. Each visual arc should overlap the previous search area by at least 10 meters in order to ensure total visual coverage of the area. Also, the search should extend as far back as the observer can see, and it should always encompass the areas of interest that were identified during the hasty search.

Maintaining Observation: Surveillance teams should repeat this cycle of hasty-then-detailed searches every 15 to 20 minutes, depending upon the terrain and specific responsibilities. When maintaining observation over time, personnel should keep their head and body movements to a minimum. They should also take special note of key terrain features and observe them closely during each visual search.

Key Terrain Features: With each consecutive visual pass over an area, personnel should take note of prominent terrain features (*positive space*) as well as any areas that may offer cover or concealment to the enemy (*negative space*). This way, observers become familiar with the terrain.

- *Positive Space: Positive space* has mass; it includes solid objects such as buildings, trees, signs, or vehicles. Personnel cannot typically see through positive space, but it naturally attracts the human eye. People are inclined to look from positive space to positive space.
- *Negative Space: Negative space* falls between positive spaces. These areas of shadow and background may be overlooked by untrained observers, which explains why good camouflage resembles negative space. Personnel must consciously observe negative spaces.
- *Other Key Features:* Personnel must remember to pay particular attention to possible anchor points, habitual areas, and natural lines of drift (these special terrain features are discussed in more detail in Unit 5). Additionally, the acronym "KOCCOA" can help warfighters and law enforcement remember the specific terrain features to observe.

Light and Shadow: Due to constantly changing clouds and the sun's positions, light is a changing factor in observation. Observers should carefully watch the changing contrast and shadows. An area that the observer first thought held no enemy may reveal an adversary when the light changes.

Rotation: To minimize fatigue and reduce the likelihood of change blindness, team members should reassign observation duty approximately every 30 minutes.

Variety of Optics: Many factors (such as distance, light level, and obstacles) affect what personnel can see with their naked eyes and/or particular optical devices. When possible, a tactical team should simultaneously employ a range of optical devices and naked-eye viewing. By using a variety of observation tools, personnel have a greater likelihood of identifying a target.

Discussion Questions:

1. How does the changing pattern of light and shadow affect an observer? Why should an observer be aware of the angle of light, especially with respect to the use of optical devices?
2. How might the natural limitations of human sensation and perception affect these observational techniques? What steps can be taken to compensate for these natural limitations?

REMEMBER!

KOCCOA (pronounced "co-CO-uh") helps you remember the high priority terrain features, these are:

K = Key terrain features
O = Observation points
C = Cover
C = Concealment
O = Obstacles
A = Avenues of approach

KEY SKILL

Efficient use of refocusing in visual scans to include both near and far objects in the scene

Instructors should emphasize the importance of keeping all parts of the viewing sector, both near and far, within one's visual field. In other words, personnel must keep their "heads on a swivel," so they are always ready to attend to critical new cues.

3005: Optical Devices

Introductory
Lecture and discussion
About 120 minutes

Terminal Objective: This module introduces trainees to the optical devices that can be used for observation in the tactical environment. Upon completion of this module, trainees will be able to describe the capabilities and limitations of the three types of optical devices, and they will be able to demonstrate accurate use of the devices.

Module Vocabulary:

- Diopter Sight

Resource DVD:

- 3005.1 – Optics Use, Video

KEY SKILL

Using organic assets to make positive identifications

Instructors should emphasize the importance of completely understanding—*and mastering use of*—one's organic optics. Trainees who do not have full mastery of their optics are at severe disadvantage.

Instructor Activities:

1. Ask trainees to bring their organic optics to class
2. Review daytime, nighttime, and thermal optics
3. Ensure all trainees know how to *fully employ* their optics
4. If necessary, break trainees into groups and review the "knob-ology" of all of the optics; encourage trainees to help each other
5. Review the Optics Use video

Synopsis of Module Topics:

Introduction: Optical devices are used in the tactical environment to enhance or increase personnel's natural ability to see. Optical devices may magnify an area, so that more detail can be seen from a greater distance, or they may enhance viewing in low-light conditions.

Daytime Devices: Day optics are unpowered devices that rely on lenses to enhance viewing. These devices are capable of viewing objects in depth and of "burning through" brush and shadow. Their primary drawback is their inability to enhance viewing in low-light conditions. Many daytime optics have a *diopter sight*, which is used to assist aiming. Personnel may not know how to adjust the diopter, which limits their devices' usability.

Nighttime Devices: Night optics operate by amplifying the ambient light in an environment. Night optics are not affected by fluctuation in ambient temperature. However, these devices are limited by their inability to see depth of field, their lack of peripheral vision, and often by their range. Additionally, night optics can cause a visual "white out" if a sudden change of ambient light occurs.

Thermal Devices: Thermal optics operate by detecting objects' infrared signatures, distinguishing an object based upon the difference of its heat signature compared to its surroundings. Thermal optical devices have the capability to detect objects at greater ranges and through obstacles. They have the greatest ability of any optical device to detect persons or vehicles in the natural environment. However, thermal optics are limited in their inability to detect friend or foe; they require the greatest power of any optical device, and they can be affected by weather.

Discussion Questions:

1. When are the best times and environments to employ the different types of optical devices?
2. Can each trainee describe and demonstrate how to stabilize their optics? When present, can trainees use their diopter sights?

3006: Optical Devices Demo (Activity)

Instructor Activities:

1. Advise trainees to bring all of their organic optics to the exercise
2. Reserve the necessary equipment and locations for the exercise
3. Position role-players at various observation points
4. Ensure trainees have full knowledge of their optics use
5. Conduct the compare-and-contrast demonstration
6. Conduct the visual-search demonstration

Synopsis of Module Topics:

Basic Use: Trainees *should* know how to use their optical devices; however, instructors should ensure all trainees know their "knob-ology."

Compare and Contrast: This demonstration helps trainees' compare their naked eyes to their optical devices. Trainees should stand approximately 100 meters from the building in which the role-players are positioned near windows or hiding in shadows (like snipers). Ask the trainees to spot the role-players with their naked eyes, with and without sunglasses; do they see any signs of shape, shadow, or silhouette? Then ask the trainees to use their optics. Discuss how the optics are designed to gather light, which allows their users to "burn into" the shadows to see the enemy. Try out the various optical devices; which of them works best, in the trainees' opinions?

Repeat this exercise several time, with the role-players in different positions and performing different actions. For instance, ask a role-player to move all the way up, into the window space. Can the trainees' see that person with their naked eyes? Then ask the role-player to slowly move back; when do the trainees lose sight of him/her? How about when the role-player moves? Does movement, even subtle movement, aid visual identification? How about when the role-player is holding a weapon; can the trainees' determine the weapon type when using their optics?

Visual Search: Pre-position a role-player out, between 50–200 meters, in an area of vegetation (or other concealment). Ensure the trainees do not see the role-player getting into position. Ask the trainees to use their visual search techniques to locate the role-player. Ensure the role-player is not moving (at least at first). Can the trainees spot the role-player? If not, if the role-player moves subtly can they spot him/her? What visual cues (e.g., movement, shine, silhouette) made the role-player most visible?

Enhanced Thermals Use: Discuss the special ways personnel can employ thermals. If available, show trainees how a warmed-up car looks through thermals. Discuss how thermals can help identify body bombers or other persons of interest; demonstrate this with role-players, if feasible.

Introductory
Demonstration
About 180 minutes

Terminal Objective: This module gives trainees the opportunity to practice basic observation skills. Instructors should reiterate lessons related to positive/negative space, scanning from right-to-left, and the different search procedures. Upon completion of this module, trainees will have functional knowledge of optics use and visual search methodologies.

PREP CHECKLIST

✓ Trainees' Optics
- Daylight Optics
- Thermals

✓ Range Location
- With Structures
- With Vegetation

✓ 3–4 Role-Players
✓ "Weapons" for Role-Players
✓ Concealment Fabrics*
✓ Instructor Comms
✓ Sunny Day

*The "concealment fabrics" support Module 3008; it is recommended that Modules 3006 and 3008 be conducted on the same day.

3007: Avoiding Observation

Introductory
Lecture and discussion
About 60 minutes

Terminal Objective: This module introduces trainees to the concept of avoiding observation by enemy forces. Upon completion of this module, the trainee will be able to discuss and demonstrate several principles of camouflage and how they can help avoid observation.

Resource DVD:

- 3007.1 – *Army Field Manual 5-20,* PDF
- 3007.2 – Avoiding Observation, Video

A BORSTAR Border Patrol Agent uses spider netting to conceal the shine from his optics and the shape of his silhouette.

Instructor Activities:

1. Review the Avoiding Observation video
2. Review the features that make items more-or-less visible
3. Hold an interactive discussion on how to minimize these features
4. Discuss how the shine from optics can be concealed
5. Discuss the 300-meter bubble

Synopsis of Module Topics:

Introduction: Personnel in reconnaissance missions should try to mask themselves and their equipment. Remember, there are almost no straight lines or regular patterns in nature, so these features will attract attention.

Why Things are Seen (Review): Camouflage and concealment are used to avoid detection. Since anything regular (e.g., a helmet, rifle, or human silhouette) stands out, they must be altered to become less noticeable. As discussed in several previous sections, the eye is drawn to shape, shine, shadow, silhouette, surface, and spacing. Personnel should disguise these features. In addition, movement rapidly draws attention; personnel attempting to avoid observation must remain as still as possible.

Shine: If personnel look towards the sun, then their optics will create a bright—highly visible—shine. However, this shine can be readily obscured by placing fabric, such as burlap or spider netting, over the optical lenses. When personnel place fabric over the front of their optics, they will need to readjust the optics' focus so that the fabric appears blurry but the surveillance targets are clear. Once complete, personnel should have no problem seeing through their concealment.

Note, objective lenses (such as binocular lenses) are designed to gather light. This allows observers to "burn through" shadows and better identify targets. Sometimes individuals tape-up their lenses, so that only a small slit is visible; while this effectively conceals the shine, it also limits the effectiveness of the optical devices.

Three-Hundred Meter Bubble: Three-hundred meters is the "magic mark," the typical range of many weapons/explosive devices. When conducting reconnaissance and surveillance, personnel should strive to remain outside the 300-meter bubble. This will enhance the safety and security of their teams.

Discussion Question:

1. Consider your own equipment. Which objects are most visible? How can they be concealed in your operational environment?

3008: Avoiding Observation (Activity)

Instructor Activities:

1. Conduct the fabric concealment demonstration
2. Conduct the back-and-forth concealment/observation practice

Introductory
Demonstration
About 120 minutes

Terminal Objective: This module gives trainees more opportunity to practice with their optical devices. This demonstration emphasizes concealment skills. Upon completion of this module, trainees will have applied knowledge of their optics use and strategies for concealment. They will be able to demonstrate effective optics use, and they will be able to list and demonstrate several concealment techniques.

Synopsis of Module Topics:

Recommendation: It is recommended that this module take place in conjunction with Module 3006 in order to minimize logistical burden. In other words, conduct Module 3006. Then conduct Module 3007, which may be carried out as an in-field discussion. Then begin this activity. The "concealment fabrics" listed in 3006's checklist are for this module.

Concealment Demonstration: Instructors should have several different concealment fabrics with them, including burlap sacks and spider netting. Pass the different materials around and ask trainees to focus their optics through them. Show trainees how to "burn through" the fabrics. Similarly, if available, ask trainees to conceal themselves behind bushes or other vegetation. They can "burn through" such vegetation just like they did with the fabrics.

Create Two Groups: Divide trainees into two groups. Make sure at least one member of the instructional staff accompanies each group, and then physically separate the groups by about 200 meters. Have the groups take turns practicing concealment or observation techniques. Let trainees sense what it "feels" like to be exposed on a hilltop or to create a visible silhouette. Then let them get a sense for how concealment feels. Similarly, ask trainees to practice remaining still. Let them get a sense for how little movement is necessary to draw the eye. Continue this back-and-forth demonstration until all of the major factors that attract vision (i.e., shape, shine, shadow, silhouette, surface, spacing, and movement) have been demonstrated and then concealed.

Left: Border Patrol Agents practice using enhanced observation techniques

POLICE
ICE
POLICE
ICE

Unit 4: The Mind of Your Quarry

4

The *Art of War* promises: "So it is said that if you know your enemies and know yourself, you can win a hundred battles without a single loss." This unit focuses on learning about one's enemies: How do they think? Why do they act in certain ways? What are they planning? After completing this unit, the trainees will be better able to see from the perspective of their quarry, their enemy, or adversary.

Suggested Prerequisites:

- CODIAC Units 1–3

Terminal Learning Objective:

- This unit provides detail on enemy decision-making. This knowledge supports four of the key training objectives: identification of critical event indicators, interpretation of human behavior cues, synthesis of ambiguous information, and proactive analysis and dynamic decision-making. More succinctly, this module will help personnel develop their *tactical cunning*, both in general as well as specifically for their own operational environment.

Enabling Learning Objectives:

- Describe combat profiling and explain how it differs from other profiling approaches
- List and define typical enemy tactics, including urban masking and soft target selection
- Describe the seven-step terrorist planning cycle and list ways to block terrorists at each step
- List several current terrorist groups and describe some of their tactics
- List all of the major terrorist groups in one's own area of operations
- List the primary tactics, techniques, and procedures of the groups in one's own area of operations

Estimated Time Allotted for Instruction: About 20 hours

Speciality Support Requirements:

- No special facilities required

4001: Introduction to Combat Profiling

Introductory
Lecture and discussion
About 60 minutes

Terminal Objective: This module introduces the concept of combat profiling, and it discusses how combat profiling applies to all people of any country or culture. Upon completion of this module, trainees will be able to clearly describe the purpose, general use, and benefits of combat profiling, as well as how it differs from racial profiling.

Module Vocabulary:

- Combat Profiling
- Explicit Knowledge
- Sustained Observation
- Tacit Knowledge

Resource DVD:

- 4001.1 – Unit 4 Intro, Video

"ALL TRIBES, ALL TEAMS, all insurgents, terrorists, and criminals are the same. They all function the same way. They all act the same way."
—*Greg Williams*

Instructor Activities:

1. Review the Unit 4 Intro video
2. Describe combat profiling; contrast it against other forms of profiling
3. Discuss how combat profiling works across cultures
4. Emphasize the importance knowing the culture and language of one's area of operations; explain why "culture is context" for profiling

Synopsis of Module Topics:

Introduction: *Combat profiling*, developed by Greg Williams, is the art of identifying behavioral cues, synthesizing them into a meaningful pattern, and then making sense of that pattern, ideally, left-of-bang. Combat profiling equips personnel with a more thorough understanding of human behavior and an ability to read the "human terrain."

Combat Profiling vs. FBI Profiling: Combat profiling is *not* FBI-style profiling. The main difference is that FBI-style profiling is reactive. In contrast, combat profiling focuses on identification of pre-event indicators and proactive profile construction, left-of-bang. FBI-style profiling also uses probabilities to estimate criminals' likely characteristics.

Combat Profiling vs. Racial Profiling: Combat profiling is *not* racial profiling. Combat profiling has nothing to do with race, sex, or the color of a person's skin. Instead it is about looking at behavioral cues and then problem-solving based upon them.

Combat Profiling and Mental File-Folders: Combat profiling requires personnel to develop mental file-folders of people and places. Once such file-folders exist, personnel can compare their current observations to their mental file-folders in order to identify anomalies from the baseline.

Developing Combat Profiling File-Folders: Combat profiling requires personnel to know the baseline, context, and relevance of their area of operations. However, personnel do *not* need to memorize these factors from a book; instead they can use *sustained observation* and gather *tacit knowledge* to create their mental file-folders.

- *Sustained Observation:* Sustained observation simply means expending conscious energy to observe an area or people, in order to develop a sense of "normal."
- *Tactic Knowledge:* When entering an area of operations, personnel can gain a sense of the baseline from their peers already operating in that region. When warfighters deploy to theater, for instance, they have transition time that they can use to glean tacit knowledge from those personnel already in-country.

Explicit vs. Tacit Knowledge: *Explicit knowledge* can be written down, transmitted, and understood by others. Explicit knowledge includes ideas that can be easily be recorded and taught, such as facts and formulas. On the other hand, tacit knowledge is gained through hands-on experience and cannot be easily written down or transmitted. Tacit knowledge is valuable because it provides context for people, places, ideas, and experiences. Effective transfer of tacit knowledge generally requires extensive personal contact and trust.

Characteristics of People Everywhere: Greg Williams, the originator of combat profiling, often says that these techniques apply to people from Kansas to Kandahar. In other words, combat profiling works everywhere. That is because it involves learning the *process* of reading people. It is not based on memorization of specific tactics or cultural differences. Williams explains the universality of combat profiling like this:

> When you go to Germany, do you have to learn the German laws of gravity? When you go to France, do you have learn French math? When you talk about the normal human body temperature, here, is it different when you get to Iraq or Afghanistan? No! People all have the same circadian rhythms. We get tired at night; we are alert in the morning; we eat basically three meals a day. We have all the same needs, all the same wants, and we respond to external simulation the same way.

Culture is Context: The only difference among people is culture, and it does exert a small influence on combat profiling. Culture is context, and learning the culture of one's area of operations is important. Additionally, understanding the language and the root meaning of words can give unique insight into a culture.

Discussion Questions:

1. Consider a "user profile" from a web site or social media outlet. What kind of information does a user profile usually store? How is an online user profile similar to the type of user profile a combat profiler creates for a person in his/her area of operations?
2. If you were entering a new operational environment for the first time, how would you establish a baseline? How can you elicit tacit knowledge from your fellow warfighters or law enforcement agents already operating in that region?
3. How do you say phrases such as "hello," "I'm sorry," "danger," or "where" in the language spoken in your operational environment? What other key phrases might be important to learn?

DID YOU KNOW?

Understanding the nuances of a language can provide great insight into a culture. Consider this example:

Salaam

When people greet each other in the Dari or Pashtu languages (which are spoken in Iraq and Afghanistan) the first person says "*as-sa-laam alai-kum*" and the second responds with "*wa alai-kum as-sa-laam.*" Both of these phrases use the root word "salaam," meaning "peace," and "salaam" also forms the root of Jeru*salem*, which translates to mean "city of peace."

"WE HAVE TO QUESTION everything we see, smell, taste, and feel. Yep, it's harder; your job has become harder in the battlespace, because you have to be the 'thinking warrior.' You have to up-armor your brain. But, at the end of the day, you'll be harder to kill, and you'll be smarter."

—*Greg Williams*

4002: Think Like the Enemy

Introductory
Lecture and discussion
About 120 minutes

Terminal Objective:

Module Vocabulary:
- Demographic
- Ideology
- Manifesto
- Psychographic
- Sticky Messages

Resource DVD:
- 4002.1 – *Root Causes of Terrorism*, PDF
- 4002.2 – Terrorists, Video
- 4002.3 – Mule Deer, Video

KEY SKILL

Taking someone else's perspective

Instructors should reemphasize the importance of using tactical cunning. Trainees need to be able to take-on someone else's perspective, putting themselves into their enemy's mindset.

Instructor Activities:

1. Review tactical cunning and describe the importance of using it
2. Discuss enemy motivations, communications, and recruitment
3. Ask trainees to read and discuss the *Root Causes of Terrorism*
4. Review the Terrorists video
5. Discuss the Mule Deer Buck metaphor

Synopsis of Module Topics:

Introduction: *Tactical cunning* means thinking like the enemy: looking at the world through their eyes, walking in their shoes, and having a day in their skin. Using tactical cunning improves a warfighter or law enforcement agent's chances of survival in a kinetic environment.

Motivation: The first step toward thinking like the enemy is to understand their motivation. Why are they radicalized? Why do they engage in radical action; why do they commit terrorism?

- *Ideology:* An *ideology* is a person's world view; ideologies are the ideals, goals, and expectations that guide actions. For this training, an ideology contains three relevant parts: culture, politics, and religion. Each of these factors may motivate radical behavior. However, according to the *US Government Counterinsurgency Guide* (2009, p.6):

  > Modern insurgencies are often more complex matrices of irregular actors with widely differing goals. At least some of the principal actors will be motivated by a form of ideology (or at least will claim to be), but that ideology will not necessarily extend across the whole insurgent network.

- *Other Motivations*: Each person may be drawn to radicalism by other individual or collective goals, including personal or political grievances, admiration for a charismatic leader (e.g., Osama bin Laden), or attraction to the money or status offered by a radical group.

Manifesto: A *manifesto* is a public declaration of an ideology. Radical groups promote themselves, in part, by glorifying their manifestos. *Root Causes of Terrorism* mentions "*sticky messages*," or simple, concrete, messages with emotional appeal and compelling storylines:

> The message that "Islam is under attack," is simple, credible (especially when bolstered by pictures of occupied lands, civilian victims of conflict), unexpected (e.g., the dissonance caused by not being a political or economic power), evokes fear (based on the implicit understanding that fear is typically the primary motivator for collective action), and contains a storyline peppered with powerful references (e.g., "crusades," "Hulagu Khan").

Iconography: Along with their manifesto, a radical group will use iconography to identify itself. Such imagery may include flags, shields, or logos; clothing; or tattoos. For example, radical Islamic terrorists use the *Shahada* (the Islamic creed) written in white on a black background as their primary iconography. Other groups, such as Mexican drug cartels, may mark their affiliation through tattoos.

Shahada, written in white on black

Terrorist Group Formation and Recruitment: The next question concerns *how* the enemy operates. Terrorists and insurgents do not typically act in isolation; instead they operate in cells and networks.

- *Terrorist Networks:* A terrorist group is bolstered by a pyramid of supporters. While the terrorists may be at the peak of the pyramid, its base is comprised of radical sympathizers and supporters. *Root Causes of Terrorism* suggests that only about 4% of radical sympathizers will become radical actors (terrorists).
- *Recruitment:* According to a RAND Corporation study on Al-Qaeda recruitment (Gerwehr & Daly, 2006), terrorist groups recruit by shaping their ideological message to suit the *demographic* and *psychographic* particulars of their audience. Potential recruits may be contacted in a number of public or private ways, such as through radio programs, web sites, school mentorship, or graffiti. Recruits tend to share common psychographic—rather than demographic—characteristics. In other words, the "attitudes, ideas, reasoning, and physical experiences of individuals weigh more heavily in their ability to resist recruitment than do such factors as their age, profession, and gender." The shared psychographic factors that make a person prone to recruitment include:
  - — High level of dissatisfaction (emotional, physical, or both)
  - — Cultural disillusionment (i.e., unfulfilled idealism)
  - — Lack of an intrinsic religious belief system or value system
  - — Some dysfunctionality in family system
  - — Some dependent personality tendencies (e.g., suggestibility)

Discussion Questions:

1. In the Mule Deer Buck metaphor, what category of battlefield actor does each animal—squirrel, rabbit, bear, coyote, mountain lion, and mule deer buck—represent?
2. What icons and ideology do groups in your operational area use?
3. Describe one new thing that you learned from reading the *Root Causes of Terrorism* report. How does it apply to your operational area?

ONLINE

Find more resources from the National Consortium for the Study of Terrorism and Responses to Terrorism at:

WWW.START.UMD.EDU

EVEN ORGANIZED CRIMINALS or criminals that are less sophisticated will have an ideology. ...Theirs is based on—*not* religion and *not* politics—their's is culture-based. ...That is what will separate a criminal organization from a terrorist.

Not every criminal is a terrorist, but all terrorists do criminal acts. The difference is that the terrorist is going to do a criminal act for the good of their cell, for the good of their ideology, where the criminal is going to do it for him, for his personal gain.

—Greg Williams

4003: Think Like the Enemy (Activity)

Intermediate
In class exercise
About 180 minutes

Terminal Objective: This module gives trainees the opportunity to explore the enemy mindset through an In class exercise on terrorist motivation, ideology, and recruiting. Upon completion of this module, trainees will have deeper understanding of terrorist mindsets; they will be able to discuss terrorist and criminal motivations in depth and describe example terrorist ideologies, manifestos, and recruitment techniques.

Resource DVD:

- 4003.1 – Think Like the Enemy Example, Video

Instructor Activities:

1. Divide trainees into their five- or six-person teams
2. Ask each team to consider the terrorists/criminals in their area
3. Ask each team to "recruit" the class for a cell/gang in their area
4. Conduct interactive after-action reviews after each demonstration

Synopsis of Module Topics:

Trainee Team Briefing: After dividing trainees into their teams, ask each team to consider the types of terrorist and criminal organizations in their own operational environments. How do these groups recruit? What are their ideologies, their manifestos? How do they create a sense of pride or reverence for, or fear of, the group?

Plan Demonstrations: Assign each team to "recruit" the rest of the class for a radical group like the ones operating in the team's area of operations. Give teams about 5–10 minutes to present their "recruiting" demonstrations to the class. In the presentations, they should use sticky messages that help convey the ideology of their group and help promote the "brand" (or identity) of the cell or gang. They should also reinforce the dissatisfaction and disillusionment that the potential recruits feel and offer those potential recruits a "better answer" through the radical group.

Give Demonstrations: After teams have prepared, each should attempt to "recruit" the class. After each 5–10 minute presentation, the instructor should lead an interactive action-action review with the class. How did each team perform? Did they convey the ideology of their group? Did they appeal to potential recruits' "hearts" and "minds"? Also, what did the trainees learn about the mindset of their opponents? How can personnel limit the appeal of such radical organizations?

Right: Trainees carry out this exercise as if they were radical Islamic terrorists; they attempt to recruit the class to join their jihad.

4004: Hide Like the Enemy

Instructor Activities:

1. Define and discuss *habitual areas* and *anchor points*
2. Discuss the significance of graffiti
3. Define and discuss *urban masking*

Introductory
Lecture and discussion
About 60 minutes

Synopsis of Module Topics:

Introduction: Terrorists and other criminals move, gather, and hide in generally predictable ways.

Habitual Areas: *Habitual areas* are public locations where anyone can gather without reservation.

Anchor Points: *Anchor points* are locations that particular individuals or groups control and where they can, therefore, gather comfortably and frequent without reservation. Individuals outside of the permitted group/sect feel uncomfortable entering such areas. For example, gang members in Los Angles may congregate at their neighborhood liquor store. This would be one of their anchor points.

- *Graffiti:* Anchor points will exhibit graffiti, a stark lack of graffiti, or other signs to identify that a particular gang or sect "owns" that location. Personnel should always pay close attention to graffiti and take note whenever that graffiti changes or is painted over by a rival group.

Hiding in Plain Sight: *Urban masking* is used to hide in plain sight. Criminals, terrorists, and insurgents actively use urban masking to help camouflage themselves in the urban environment—attempting to blend into the baseline around them. For instance, a terrorist might hide an IED in car along the side of the road, and then put the car's hood up so that it appears to be broken down. It is also common practice, from Mexico to Afghanistan, to attempt to look like the "good guys," dressing in the uniforms of police, governmental military, or international forces.

Discussion Questions:

1. In the Juba sniper incident, how did the Juba sniper team use urban masking? Why did they use a vehicle with a "taxi" sign? Where do you think they parked their car?
2. How do the physiological and cognitive limitations discussed in Units 2 and 3 help enemies get away with the use of urban masking? How might you overcome these natural limitations?
3. How does urban masking relate to humans' natural propensity to enter a state of *denial* when an anomaly occurs?

Terminal Objective: This module introduces trainees to concepts related to enemy movement, gathering locations, and hiding techniques. Upon completion of this module, trainees will be able to define and describe the importance of habitual areas, anchor points, and urban masking.

Module Vocabulary:

- Anchor Point
- Habitual Area
- Urban Masking

Resource DVD:

- 4004.1 – Hiding in Plain Sight, Video

CRIMINALS, TERRORISTS, AND INSURGENTS, they hide in plain sight, do things right in front of you, and your brain is not geared to catch it.

—*Greg Williams*

4005: Plan Like the Enemy

Introductory
Lecture and discussion
About 120 minutes

Terminal Objective: This module introduces the seven-step terrorist planning cycle. Upon completion of this module, trainees will be able to recall and discuss the seven phases of the cycle.

Module Vocabulary:

- Hard Target
- Seven-Step Terrorist Planning Cycle
- Soft Target

Resource DVD:

- 4005.1 – *Military Guide to Terrorism*, pages 117–121
- 4005.2 – Terrorist Planning Cycle, Video

Instructor Activities:

1. Ask students to look over the *Military Guide to Terrorism*
2. Review the Terrorist Planning Cycle video
3. Describe each phase of the *seven-step terrorist planning cycle*
4. Define and give examples of *hard* and *soft targets*

Synopsis of Module Topics:

Introduction: The *Military Guide to Terrorism* lists several different definitions for terrorism, but in general terrorism can be described as political, psychological, coercive, dynamic, and deliberate. While there is no universal model for terrorist operations, most effective terrorists execute the *seven-step terrorist planning cycle*. Enemies may not knowingly follow this cycle; however, learning it can help personnel exercise their tactical cunning and think like enemy terrorists.

[1] Broad Target Selection: Step 1 is broad target selection. During this phase, diverse sources provide the terrorist cell or decision-maker with information on and suggestions for potential targets. Targets may be identified based upon the impact of a strike against them, their general location, their vulnerabilities, or the organization's ability to mount an operation against them. In general, terrorists seek targets that offer a maximum payoff for a minimum risk (that is, cost versus benefit).

- *Hard vs. Soft Targets:* Often, terrorist operations try to avoid targets of strength (*hard targets*) and instead select targets that have minimal defensive capabilities (*soft targets*).

[2] Intelligence Gathering and Surveillance: Once a broad set of targets has been compiled, the terrorist organization begins gathering intelligence on each target, in order to identify the optimal one. This process may take just few days, or it may require several years. Gathered intelligence will include practices, procedures, and routines of the location; residence and workplace information; transportation and routes of travel; and security.

[3] Specific Target Selection: Using the intelligence gathered in step 2, the terrorist organization will next select a specific target based upon:

- What impacts will be felt, beyond the immediate victims
- Whether significant media attention will be garnered
- Whether it sends a statement to the target audience
- Whether its impact is consistent with group objectives
- Whether it demonstrates the group's capabilities
- Whether it offers a good cost-to-benefit ratio

[4] **Attack Surveillance and Planning:** During step 4, the terrorist group gathers specific information on the target's patterns, from a short-term perspective. The attack team will also confirm intelligence gathered from previous efforts. The attack team will focus on:

- Studying the target's security
- Detailing preparatory operations
- Identifying necessary special operatives (if needed)
- Selecting a target area base of operations
- Identifying escape routes
- Identifying necessary weapons

"TERRORISM'S EFFECTS are not necessarily aimed at the victims of terrorist violence. ...Victims are simply the first medium that transmits the psychological impact to the larger target audience."

—*Military Guide to Terrorism, page 15*

[5] **Rehearsal:** In step 5, the terrorists conduct rehearsals in order to improve their odds of success, test their plan, and explore alternative strategies. The rehearsals may also test the security responses of their target. Rehearsals may be conducted by terrorist agents themselves, or intelligence may be gathered by observing the behavior of unaware individuals. Rehearsals will include:

- Weapons training
- Final preparatory checks
- Deployment to target area
- Actions on objective
- Escape and evasion routes

[6] **Actions on the Objective:** In phase 6 actions on the objective are carried out. This is the actual terrorist attack executed by the terrorists.

- *Target of Opportunity:* Criminals and terrorists often attack *targets of opportunity*. These are soft targets that, in short, are "in the wrong place, at the wrong time."

[7] **Escape and Exploitation:** Step 7 involves escape (if the attack was not planned as a suicide attack) and, importantly, exploitation. The psychological impact on the target population far exceeds any military value of the target. In many recent conflicts, exploitation often involves posting photos and videos of the event on the Internet or sending these media to news organizations who play them on air.

Discussion Questions:

1. At what point in the planning cycle are terroristic acts most vulnerable to counter-terrorist efforts?
2. Describe some soft and hard targets in your area of operations?

4006: Dangerous Adversaries

Introductory
Lecture and discussion
About 120 minutes

Terminal Objective: This module introduces some of the terrorist groups and other "dangerous adversaries" that personnel may face. Upon completion of this module, trainees will be able to describe specific examples of terrorist operations, and they will be able to discuss specific terrorist tactics.

Module Vocabulary:

- Second-Order Effects
- Third-Order Effects

Resource DVD:

- 4006.1 – *A Military Guide to Terrorism*
  — pages 107–116,
  — pages 123–131
- 4006.2 – *Al-Qaeda*, PDF
- 4006.3 – *Mexico Cartels*, PDF
- 4006.4 – *Iran Syria*, PDF
- 4006.4 – *Taliban*, PDF
- 4006.4 – *Tamil Tigers*, PDF

Instructor Activities:

1. Ask trainees to review the terrorist groups; use the resources provided
2. Discuss specific terrorist groups, their tactics, and philosophies
3. Discuss *second-* and *third-order effects*

Synopsis of Module Topics:

Introduction: A substantial number of criminal organizations engage in terrorist and irregular warfare activities. Nearly all military and law enforcement personnel will have to confront such "dangerous adversaries" at some point. This module discusses some specific enemies' tactics, techniques, and procedures.

Adversaries: While not comprehensive, the list below highlights some of the most dangerous terrorist organizations currently in operation:

- Al-Qaeda ("The Base")
- La Eme (Mexican Mafia)
- Gulf Cartel (Cártel del Golfo)
- Hamas ("Islamic Resistance Movement")
- Hezbollah ("Party of God")
- Juárez Cartel (Cártel de Juárez)
- Michoacan Cartel (La Familia Michoacana)
- MS13 (Mara Salvatrucha)
- Taliban ("Students")
- Tamil Tigers (Liberation Tigers of Tamil Eelam; LTTE)

Using the resources provided, as well as additional independent searches, conduct an interactive discussion on the motivation, ideology, tactics, and past operations of each of these organizations.

Second- and Third-Order Effects: A primary effect is the immediate impact of an action. For instance, a primary (or first-order) effect of the World Trade Center attack was the destruction of the towers and death of nearly 3000 individuals. A *second-order effect* is the reaction to a first-order effect. In the World Trade Center example, one second-order effect was the decline in US air travel following the attack. A *third-order effect*, in turn, is the response to a second-order effect. Again, using the World Trade Center example, a third-order effect, in response to the decline in air travel, was that airlines struggled with financial problems, forcing many companies to seek new income strategies or off-load costs in novel ways. Terrorist groups seek to maximize the impact of their acts, inflicting substantial second- and third-order damage.

4007: Plan Like the Enemy (Activity)

Intermediate
In class exercise
About 3½ hours

Instructor Activities:

1. Divide trainees into their five- or six-person teams
2. Assign each group a terrorist organization to represent
3. Ask each group to choose a real, local target and plan an attack
4. Ask groups to present their plans and then discuss them, as a class

Terminal Objective: This module gives trainees the opportunity to apply their knowledge of terrorist organizations and the seven-step terrorist planning cycle. Upon completion of this module, trainees will have a deeper understanding of terrorist operations, and they will be able to discuss terrorist tactics and planning in detail.

Resource DVD:

- 4007.1 – Terrorist Trainee Exercise - Part 1, Video
- 4007.2 – Terrorist Trainee Exercise - Part 2, Video

Synopsis of Module Topics:

Preparation: After dividing trainees into their teams, assign a real-world terrorist organization to each team. The teams should research this organization in more detail and approach this planning activity from that organization's perspective.

Instructions to Trainees: Trainees will need to seek out a real-world target for their fictional terrorist attack. Trainees must plan their attack using the seven-step terrorist planning cycle. In addition to detailing each step of the planning cycle, teams should research and be prepared to answer questions related to the following:

- What is the terrorist group's ideology and motivation?
- How will the cell acquire weapons and other necessary equipment?
- If necessary, how will the cell finance the operation?
- How will the cell train its operatives?
- If necessary, how will the cell employ urban masking?
- What are the second- and third-order effects of the operation?

Team Planning: Give trainees sufficient time to plan and research their attacks. Ask them to create concept-diagrams of their operations (these will help them consider all angles of the problem as well as how the different facets affect one another). Also, encourage trainees to make demonstration aids. For instance, they may take videos or photos of their target, or they may create a line drawing of their target's layout.

Presentation: Ask teams to present their plans to the class. Each team should be allotted about 10–15 minutes. Following each presentation, conduct a short after-action review, prompting trainees to provide extra detail, particularly related to each of the seven steps and the questions listed above. Following all of the presentations, as a class select the optimal target. Discuss why this target, and its attack plan, offer the highest cost-to-benefit ratio for the terrorist group. Then discuss potential strategies for stopping such an attack, the pre-event indicators that the terrorists may reveal, and other vulnerabilities in the operation that personnel could exploit in order to block the terrorist action.

A trainee draws a concept-diagram of his team's terrorist operation

4008: Your Area of Operations (Activity)

Intermediate
In class exercise
About 5 hours

Terminal Objective: For this module, trainees will independency research additional criminal organizations present in their own operational environments. Trainees will complete worksheets that summarize their adversaries, and they will present synopses of these groups to the class.

Resource DVD:

- 4008.1 – Worksheet

Instructor Activities:

1. Divide trainees into their five- or six-person teams
2. Ask teams to research criminal groups operating in their own area
3. Ask teams to complete the included template
4. Ask teams to present short overviews of the organizations operating in their area; pass out copies of the template to the other trainees

Synopsis of Module Topics:

Preparation: Ask each team to investigate the active criminal and terrorist organizations in their operational environment. Give the trainees sufficient time to research these groups. Some regions might be home to only one criminal organizations, but other regions may be used by several criminal groups.

Presentation: Ask trainees to complete the criminal group worksheet for each organization operating in their region. This worksheet is provided in on the resource DVD; an example is given on the next page. After trainees have researched their region and completed worksheet templates for each criminal group in their area, reconvene the class. Ask each team to give a short presentation on the criminal organizations they studied; make copies of their worksheets to distribute to the class.

Right: Trainees research terrorist groups and prepare for a class presentation

Created by: John Doe Date: 7/2010

Organization: La Eme - Mexican Mafia (Surenos)

Other Names: El Eme, Mexican Mafia, La Mafia, Mafioso

Territory: Southern California; prison system

Local Area of Operation: East Lost Angeles

Leaders: Not certain

SYMBOL/ICON

Local Leaders: Not certain, but hierarchy structure

Ethnicity: Mexican-American men

Size: About 1500 members

Allies: Aryan Brotherhood, Mexikanemi, Nazi Lowriders

Rivals: Nuestra (North) and Sureno (South) are offshoots at war with each other

Criminal Activities: Extorting drug distributors; dealing drugs (coke, heroin, meth, pot), on the streets and in prison

Recruiting: In prison

Finance: Drug distribution

Ideology: Make money; anti-Black; anti-homosexual

Icons/Symbols Used: '13' (the thirteenth letter of the alphabet, 'M'), "MM," Sur, XIII, X3, 13, or 3-dots. They use the color blue and a black hand

Typical Tactics and Operations: Direct ties with Mexican drug sources; controls most of the drug traffic in California prisons. Has a reputation for violence and terrorism.

EXAMPLE

Unit 5: Reading the Human Terrain

In the contemporary irregular environment, personnel can no longer recognize the enemy by their uniforms, flags, sex, or age. Instead of facing a recognizable foe with recognizable rules of warfare, today's personnel must examine almost every person they encounter and determine their unique threat levels. This unit introduces the *six domains of profiling*, which can help personnel interpret the cues they observe from people and their environments.

Suggested Prerequisites:

- CODIAC Units 1–4

Terminal Learning Objective:

- This unit introduces techniques for interpreting human behavior cues. At the conclusion of this unit, trainees will be able to describe the six domains of profiling in detail, and they will be able to demonstrate the domains' use on a limited scale, such as interpreting photographs.

Enabling Learning Objectives:

- Explain how the six domains support combat profiling
- Describe biometrics; list several biometric cues and their interpretations
- Describe kinesics; list several kinesic cues and their interpretations
- Describe proxemics; list several proxemic cues and their interpretations
- Describe geographics; list several geographics cues and their interpretations
- Describe atmospherics; list several atmospheric cues and their interpretations
- Discuss the importance of symposium and iconography
- Discuss heuristics and explain how heuristic matches are made
- List and interpret the relevant icons and symbols in one's own area of operations

Estimated Time Allotted for Instruction: 18 hours

Speciality Support Requirements:

- This module requires interpretation of visual imagery; instructors should have access to digital projectors or hard-copy print outs of the materials.

5001: Biometrics

Introductory
Lecture and discussion
About 120 minutes

Terminal Objective: This module introduces trainees to the concept of biometrics. Upon completion of this module, trainees will be able to discuss biometrics, list several biometric cues, and describe how biometric indicators can support various missions.

Module Vocabulary:

- Adrenaline
- Biometrics
- Blushing
- Endorphins
- Flushing
- Histamines
- Nystagmus
- Pupil Dilation

Resource DVD:

- 5001.1 – Biometrics, Video

Instructor Activities:

1. Define and explain the *biometrics* domain
2. Describe how to apply biometrics in the battlespace
3. Review the Biometrics video
4. Discuss specific biometric cues and what they may mean

Synopsis of Module Topics:

Introduction: The six domains are divided into two sub-categories, personal cues and environmental cues. Personal cues involve the behavior, movement, and body language of individuals, or more specifically, their *biometrics* and *kinesics*.

Biometrics: The biometrics domain involves interpretation of the autonomic physiological reactions that all humans naturally display. Biometric cues are impossible to hide, cannot be faked, and are not culturally dependent. Biometric cues are particularly useful in interrogation and questioning situations; because the clues are less noticeable with distance, observers must normally be close to the target. However, some biometric indicators can been seen through optical devices.

Sample Biometric Cues: Many biometrics cues are initiated by the secretion of hormones, which also initiate the fight-flight-or-freeze response:

- *Blushing: Blushing* occurs when a person's skin colors red as a result of an emotional response, such as embarrassment or shame. This kind of blushing is caused by *adrenaline,* which is released when a person experiences fight-flight-or-freeze. Adrenaline causes the blood vessels in a person's face, neck, and ears to dilate, which allows more blood to flow and gives the skin a ruddy appearance.
- *Flushing:* Humans' cheeks can become ruddy in other ways, too; however, only emotional responses, such as embarrassment or shame, cause blushing triggered by adrenaline. *Flushing,* which is more pronounced than blushing and generally covers a greater body area, is caused by the release of different hormones, such as *histamines.* Drinking alcohol, becoming sexually aroused, or abruptly ceasing physical activity can trigger flushing. Anger can also cause flushing.
- *Flared Nostrils, Showing Lower Teeth:* Flaring nostrils or showing lower teeth are instinctual indicators of fight-flight-or-freeze; they often indicate anger and aggression.
- *Paleness:* The addition or absence of *adrenaline* can also cause a person to turn pale. This can be caused by emotional shock, stress, or stimulant use.

Some biometric cues may be instigated by a wide variety of triggers:

- *Pupil Dilation: Pupil dilation* refers to the size of the pupil, or black-part, of the eye. Pupils enlarge in response to low-light conditions, drug use, or interest (including sexual attraction). When someone sees something (or someone) to which they are attracted, his/her pupil enlarges. Slow dilation may also indicate drug use.
- *Sweating:* Excessive sweating can indicate nervousness or nausea. For instance, suicide bombers often exhibit excessive sweating. Salt stains on clothing (from perspiration) or clammy skin may similarly denote nervousness or nausea.
- *Thermal Signature:* Each person has a natural heat signature. When a person's body temperature rises, either from physical exertion or emotion (anger, arousal), his/her heat signature will change. This change can be identified through thermal optics, or if personnel are physically close, the heat can be directly felt. For instance, if a fugitive runs from police and then joins a large standing crowd, the police can identify their target by looking for his high heat signature.

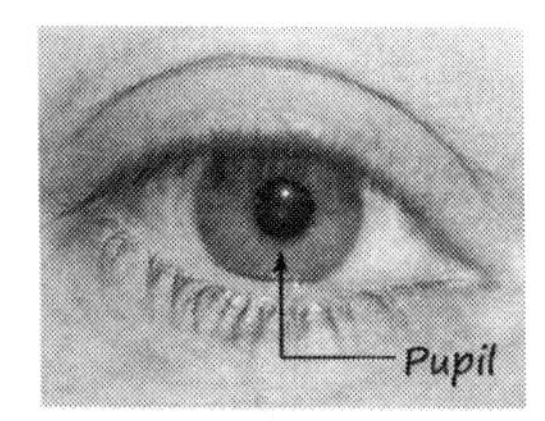

Pupil dilation

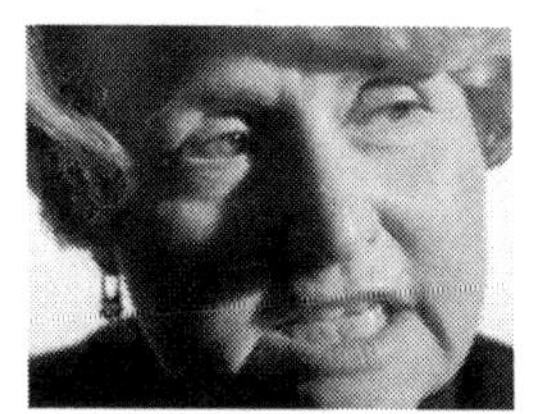

Revealing lower teeth

Other cues tend to be triggered by drugs or physical trauma:

- *Nystagmus:* Involuntary eye movement, or *nystagmus*, is typically caused by ingestion of alcohol or drugs.
- *Bloodshot Eyes:* Vasodilation of the eye, or bloodshot eyes, occurs when the blood vessels in the eye dilate. Drinking alcohol or ingesting certain drugs can cause bloodshot eyes.
- *Bruising or Broken Capillaries:* Physical trauma can leave telltale signs. For instance, the recoil of a weapon may leave a mark on a person's shoulder or eye, or the straps that attach drug bundles to smugglers' backs may dig-in to their shoulders, leaving marks.

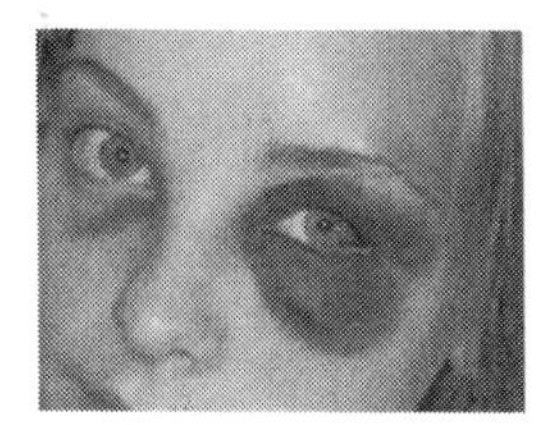

Black eye (bruise)

Finally, a variety of cues are primarily set-off by instinctual triggers:

- *Lactation:* Females lactate in response to babies. A new mother will begin lactating if she hears a baby cry, even if it is not her own.

Biometric Cues and Urban Masking: Often, individuals will employ urban masking to cover their biometric cues. For instance, lactating mothers may wear nursing bras to disguise unwanted lactation, or marijuana users may employ eye-drops to disguise their bloodshot eyes.

Discussion Question:

1. What examples of these biometric cues have you seen in real life? What cues do people exhibit under stress? Fatigue? Anger?

5002: Kinesics

Introductory
Lecture and discussion
About 120 minutes

Terminal Objective: This module introduces trainees to the concept of kinesics. Upon completion of this module, trainees will be able to discuss kinesics, list several kinesic cues, and describe how kinesic indicators can support various missions.

Module Vocabulary:

- Kinesics
- Lecturing

Resource DVD:

- 5002.1 – Kinesics, Video
- 5002.2 – Kinesics Presentation, PPT

Instructor Activities:

1. Define and explain the *kinesics* domain
2. Describe how to apply kinesics in the battlespace
3. Review the Kinesics video
4. Discuss specific kinesic cues and what they may mean; use the Kinesic presentation to support this discussion

Synopsis of Module Topics:

Introduction: People give and read thousands of nonverbal messages everyday, and people react emotionally to nonverbal messages without consciously understanding them. *Kinesics* is the interpretation of these nonverbal cues, which include body language, gestures, facial expressions, grooming habits, and positioning of the body in space.

Kinesics: Kinesic cues may be learned, innate, or a mixture of the two. For instance, learned kinesics include eye-winks and thumbs-up signs; whereas, blinking and fist-balling are innate kinesics. Laughing and crying are examples of mixed (both innate and learned) kinesics; they originate in as innate actions, but cultural rules later shape their expression.

Sample Kinesic Cues: A variety of body movements can be interpreted:

- *Telling the Truth:* When people slap the front of their heads, they are touching their prefrontal cortex (or the "thinking" part of the brain). This often indicates a person is genuinely recalling information; similarly, when individuals look up or horizontally to the left or right when remembering, this often indicates that they are recalling real facts. Individuals who look up tend to focus on visual recall, while those who look horizontally left or right focus on auditory recall.
- *Attempting to Deceive:* Individuals attempting to deceive may scratch their heads or rub the backs of the heads. These behaviors are caused by sweat emerging on a person's scalp (from anxiousness, as discussed in the biometrics module) or the feeling that his/her hair is standing-on-end (also a biometric reaction to lying).

  Closing one's eyes indicates that a person wants to (metaphorically) forget or ignore the topic of discussion or object in front of them. For instance, a person who is lying may close his/her eyes and even cover them with a hand, to "hide" from the person he/she is deceiving. Individuals may often cover their mouth when lying; hiding the lie.
- *Nervousness:* Wringing hands or rubbing palms together slowly may signify nefarious intent or nervous energy, while rubbing palms together vigorously may signify anticipation of a successful outcome.

Body placement may imply certain thoughts or feelings:

- *Power and Authority:* Placing hands on one's hips ("pointing" at one's genitals) indicates a sense of power. Individuals in roles of authority (or wishing to be in authority) may use this cue. In contrast, when people feels threatened, they tend to hide their genitals.
- *Attention:* The direction a persons' toes points indicates where that individual is focusing his/her attention. If the toes point toward another person, this indicates the individual is paying attention to his/her companion; whereas, if the toes point towards the door, then he/she is ready to leave.

Kinesics also includes facial expressions:

- *Smiles:* Genuine smiles are uneven; only "fake" smiles are symmetrical. Genuine smiles are also accompanied by wrinkling throughout the face and around the eyes.

Verbal pauses or unnecessary expressions typically indicate deception:

- *Pregnant Pauses:* If someone pauses before answering, particularly simple or straightforward questions, this may indicate that he/she is lying (and attempting to craft the lie). Along the same lines, liars may repeat the question or use other delaying tactics to give themselves time to craft a lie.
- *Unnecessary Detail:* Individuals may provide unnecessary detail in response to a question. This also may indicate a person is trying to craft a lie and make it sound believable through detail.

Cultural Kinesics: A small set of kinesic cues may be culturally dependent. For instance, in the Middle East showing someone the bottom of one's foot is an insult and may mean that the individual is disgusted. In some cultures, it is respectful to avoid eye-contact, while others show respect by making eye-contact. When entering a new area of operations it is important to identify the cultural kinesics of the region. Additionally, remember that people act differently when they are under stress or during critical incidents.

Crossing arms creates a barrier

Liars often rub or close their eyes

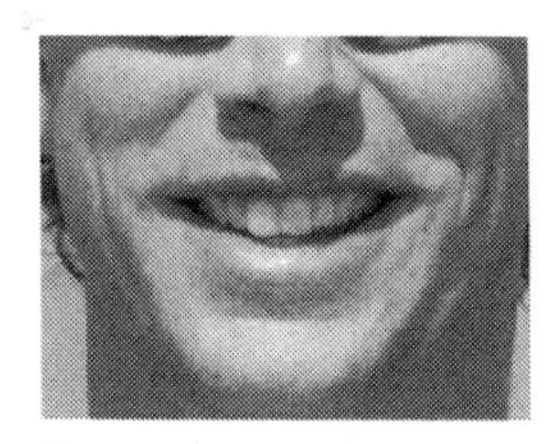

Genuine smiles are uneven

Discussion Questions:

1. How can kinesics help support your own missions?
2. What other kinesics cues have you witnessed? What cues do people exhibit under stress? Fatigue? Anger?
3. What kinesic cues do people display when they are creating a barrier? What does crossing one's arms imply?

5003: Proxemics

Introductory
Lecture and discussion
About 120 minutes

Terminal Objective: This module introduces trainees to the concept of proxemics. Upon completion of this module, trainees will be able to discuss proxemics, list several proxemic cues, and describe how to proxemic indicators can support various missions.

Module Vocabulary:

- Adoration
- Direction
- Entourage
- Mimicry
- Proxemic Pull/Push
- Proxemics
- Save/Lose Face

Resource DVD:

- 5003.1 – Proxemics, Video

REMEMBER!

Proximity negates skill. The closer an enemy is to someone, the less skill he/she needs. By observing adversaries from a greater distance, personnel increase the distance between themselves and the potential threats. This distance gives personnel more time to make decisions.

Instructor Activities:

1. Define and explain the *proxemics* domain
2. Describe how to apply proxemics in the battlespace
3. Review the Proxemics video
4. Discuss specific proxemic cues and what they may mean

Synopsis of Module Topics:

Introduction: As mentioned previously, the six domains are divided into two sub-categories, personal cues and environmental cues. Environmental cues include the distance between people or things, the physical features of a space, how individuals interact with a space, and the overall "feel" of an area. In other words, environmental cues include the *proxemics*, *geographics*, and *atmospherics* domains.

Proxemics: *Proxemics* involve the interpretations of spatial relationships within the context of cultural norms, tactical considerations, and psycho-social factors.

Average Distances: The relative distance between individuals or groups of people can indicate their familiarity or comfort with each other:

- *Intimate Distance:* Embracing, touching, or whispering
  - Closer Societies: Less than 6 inches
  - More Distant Societies: 6-18 inches
- *Personal Distance:* Interaction among close friends or family
  - Closer Societies: 1.5 to 2.5 feet
  - More Distant Societies:2.5 to 4.5 feet
- *Social Distance:* Interaction among acquaintances
  - Closer Societies: 4 to 7 feet
  - More Distant Societies: 7 to 12 feet
- *Public Distance*: Interaction for public speaking
  - Closer Societies: 12 to 25 feet
  - More Distant Societies: Greater than 25 feet

These spatial relationships are affected by cultural differences. European and North American cultures, for example, tend toward more distant proximity, and Arab and Latin cultures tend towards closer proximity.

Proxemic Push/Pull: *Proxemic push* occurs when a person or group uses body language to create distance to another person. In contrast, *proxemic pull* occurs people use body language to invite others toward them. For instance, slightly turning one's back on another creates a proxemic push.

Identifying Leaders Through Proxemics: Proxemic indicators can reveal the leader of a group, who may be a High Value Target (HVT) or High Value Individual (HVI):

- *Direction: Direction* occurs when one member of a group consistently gives orders to the other members. Leaders may give overt direction, but they also communicate direction more discretely. For instance, a leader may use the practice of "once removed," where they only give orders directly to their deputy, who then interacts with others. This nonverbal cue seems to elevate the leader to a superior status.
- *Entourage:* An *entourage* is one or more people following (i.e., in a beta position to) another person. An entourage will show subordination or submissive behavior to the leader; for instance, an entourage member may glance over at the leader to visually confirm that his/her actions are appropriate (e.g., looking for an eyebrow flash or slight nod).
- *Mimicry: Mimicry* occurs when a group or individual mirrors the actions of another. For instance, imagine two people are chatting and the first crosses his legs and leans forward, if the second man is positively engaged in the conversation, he will mimic this cue. That is, he will also cross his legs and lean forward. Mimicry occurs when people are in agreement or engaged with one another; additionally, subordinates will mimic their leaders' body language.
- *Adoration: Adoration* is the outward sign of affection towards an individual by another individual or group. Adoration cues may be positive or negative (i.e., an outward lack of adoration). Subordinates will generally show positive adoration towards their leaders; example cues include slightly bowing one's head when greeting a leader, standing when a leader arrives, or opening a door for a leader.

Proxemic space may mean dissatisfaction

Happy couples display tight proxemics

School children display mimicry

Reciprocation: Personnel who wish to build trust and rapport with another person must reciprocate proxemic cues appropriately. For instance, in the Middle East, a village representative may greet warfighters by hugging and kissing them. If a warfighter fails to appropriately reciprocate the gesture, and instead displays discomfort and unease, the village representative will *lose face*. In other words, he will have been publicly embarrassed.

Discussion Questions:

1. What proxemic cues will indicate whether a person is new or unfamiliar to a group? Who the group admires? Who the group dislikes or perceives as a threat?
2. Can you think of an example of saving or losing face? How did (or could have) someone helped the person *save face*?

5004: Geographics

Introductory
Lecture and discussion
About 120 minutes

Terminal Objective: This module introduces trainees to the concept of geographics. Upon completion of this module, trainees will be able to discuss geographics, list several geographic cues, and describe how to geographic indicators can support various missions.

Module Vocabulary:

- Channel
- Geographics
- Geographic Profiles
- Natural Lines of Drift

Resource DVD:

- 5004.1 – Geographics, Video
- 5004.2 – IED, Video

"PEOPLE FROM AN ANCHOR POINT will prey on those who normally go to a habitual area, but nobody from a habitual area will mess with an anchor point."

—*Greg Williams*

Instructor Activities:

1. Define and explain the *geographics* domain
2. Review the concepts of *anchor points* and *habitual areas*
3. Describe how to apply geographics in the battlespace
4. Review the Geographics video
5. Discuss specific geographic cues and what they may mean

Synopsis of Module Topics:

Introduction: *Geographics* is the study of the physical terrain (both natural and man-made), and the interpretation of the relationship between people and their physical surroundings.

Review: In Unit 4, trainees were introduced to the concepts of *anchor points* and *habitual areas*. They were also taught to look for graffiti and similar symbols that belie the presence of an anchor point. These concepts relate to the geographics domain. Remember, habitual areas are public gathering spots that anyone can enter; in contrast, anchor points are "owned" by a particular group or organization, and only affiliates of that clique are permitted to freely enter. Anchor points are inherently restrictive, and they often have overtly controlled entries and exits.

Micro Anchor Points: Some locations may include both habitual areas and anchor points. For instance, a mosque may be a habitual area where everyone from a village worships, but its back rooms may be used by insurgents to plan terrorist attacks, making those rooms an anchor point for the terrorist cell. Similarly, a classroom may be a habitual area, but over time each student "claims" his/her seat. In other words, students begins to sit in the same seat each class meeting, and their respective seats becomes comfortable to them. Each seat therefore becomes each student's anchor point. Houses, and even cars, may contain personal anchor points, as Greg Williams explains:

> When you are conducting a search of a house, every single house, hovel, shed in Afghanistan will have a weapons cache. They'll have a cache site for the books that they are not supposed to have, for food they are not supposed to have, and for weapons they are not supposed to have. It's the same as executing a warrant in East Los Angeles. When I go into a place, I say... "look, I don't want to toss your whole house, you know what I'm here for..." And I watch the person. I look at the house. I figure out the areas of the house that are habitual areas and then the areas that are personalized to my suspect—his geographic anchor points. ...Then I'll kind of guide him around the house, and I'll watch him as he looks around. Then—boom—he'll look at the dresser. [And I'll know where he keeps his cache.]

Familiarity: How individuals interact with the natural and man-made terrain around them can indicate their level of familiarity with the location. People familiar with an area will act, walk, and drive differently.

- *Comfort:* When in an unfamiliar area, people are attracted to places that remind them of their anchor points. For instance, criminals often commit crimes (or stash evidence) in areas similar to the type of region in which they currently live or in which they were raised.

Rubble creates a channel in Afghanistan

Natural Lines of Drift: *Natural lines of drift* are pathways used by people who are familiar with that area. Natural lines of drift are comfortable pathways; that is, they are predicable, routine, or habitually used paths. Natural lines of drift always follow paths of least resistance. In other words, they are shortcuts, and they generally connect habitual areas and/or anchor points to one another.

Indian women gather at an anchor point

Channel: *Channels* funnel or guide the movement of people. Channels may be created intentionally, such as by a mall owner attempting to guide the flow of pedestrians, or unintentionally, such by the natural formation of a landscape. In irregular conflicts, adversaries may use debris and obstacles to create channels; these pathways then guide the flow of traffic (such as military convoys) in specific directions (such as towards a buried IED). Military and police personnel also create channels. Every Vehicle Check Point (VCP) is a channel; instead of using debris, personnel may use traffic cones or police tape, but the channeling effect is the same. The danger created by the channel—that is, that many individuals become tightly packed together—may also be the same. This is one reason why VCPs are often targeted by suicide bombers or VBIEDs.

Soldiers follow a natural line of drift

Geographic Profiles: Certain activities require certain landscapes. For instance, a helicopter can only land on flat terrain and not all locations make suitable observation points. Similarly, people may be drawn to certain geographic profiles. For instance, if an IED-placer has been successful using a certain type of terrain, over-and-over again, then he is likely to continue planing IEDs in similar types of locations.

Discussion Questions:

1. What geographic profiles are used in your operational environment? In what kind of areas do certain gangs or sects typically gather or commit their crimes?

2. Anchor points can be mobile, because an anchor point is based upon the activities—not the physical terrain—in an area. What is an example of a mobile anchor point?

REMEMBER!

Humans, by nature, are *lazy*. They follow predictable patterns of behavior and take paths of least resistance.

5005: Atmospherics

Introductory
Lecture and discussion
About 120 minutes

Terminal Objective: This module introduces trainees to the concept of atmospherics. Upon completion of this module, trainees will be able to discuss atmospherics, list several atmospheric cues, and describe how atmospheric indicators can support various missions.

Module Vocabulary:

- Atmospherics
- Atmospheric Shift
- Rubble-ing

Resource DVD:

- 5005.1 – Atmospherics, Video
- 5005.2 – Graffiti, Video
- 5005.3 – *Graffiti*, PDF

"ATMOSPHERICS alone are the most powerful indicator of what's about to happen, and its the most left-of-bang."

—*Greg Williams*

Instructor Activities:

1. Define and explain the *atmospherics* domain
2. Ask trainees to look over the *Graffiti* article
3. Describe how to apply atmospherics in the battlespace
4. Review the Atmospherics video
5. Discuss specific atmospheric cues and what they may mean
6. Emphasize the importance of graffiti and discuss specific cues

Synopsis of Module Topics:

Introduction: *Atmospherics* describes the interpretation of the environmental mood. This includes how a place looks, sounds, tastes, feels, and smells. Every baseline has an atmosphere, as does every vehicle, crowd, or event. Changes in the atmosphere may be subtle pre-event indicators.

Atmospherics Example: An American visits London at the end of April and spends several days in a vibrant urban atmosphere, filled with commuters and shoppers. Then, one morning, the American finds the streets suddenly empty and the shopkeepers boarding up their stores. These are atmospheric cues, and although the visitor does not know what they imply, they are obviously not good. Later, the American discovers that it is the first of May, the May Day holiday, which is celebrated in England by anarchists who riot and loot.

Atmospherics and Context: Atmospheric cues are context-dependent. For instance, if a police officer sees a nice car, in a nice neighborhood, driving slowly while the driver looks around, that officer may think the driver is looking for a particular house. However, if the same law enforcement agent sees the same car, driving slowly with the driver scanning the area in a bad part of town, the officer may be suspicious.

Atmospheric Shift: Atmospheric cues may appear spontaneously, often in response to the presence of warfighters or law enforcement personnel. These sudden changes are called *atmospheric shifts*, and such anomalies generally indicate danger. The most obvious cues are the sudden absence of normal routines, patterns, and attitudes of the local populace or the presence of abnormal activity. Examples include:

- Observers suddenly emerging in windows or on rooftops
- Unusual absence of pedestrian traffic
- Stores, markets, or street vendors closed suddenly
- Unfamiliar vehicles or groups within the area
- Civilian workers failing to appear at military bases
- People suddenly engaging in private activities in public places

Graffiti: As previously discussed graffiti (or a stark lack of graffiti) can be a powerful indicator. Graffiti is often used as a "signpost" for either public or private communication. Private graffiti, for instance, may include a sloppy tag on the back of a street sign that points members of a particular gang towards the gang's anchor point. Public graffiti may include a large sophisticated tag, painted on the side of a prominent building.

- *Crossing Out Graffiti:* Rivals may cross-out or mark over each others' graffiti. Marking over someone else's graffiti may indicate a threat, a claim to territory, or a statement of superiority.

A shop in Afghanistan sells fuel oil

Trash and Rubble: Like graffiti, trash, rubble, or other debris may fill a dangerous area. Not only can such "*rubble-ing*" camouflage people, paths, or activities (urban masking), but it can also create channels. An area becomes debris-strewn when criminal or terrorist groups want it to become cluttered or when the people living at the location are unmotivated or discouraged to clean it.

Rubble blocks a street in Afghanistan

Suspicious Items: Personnel must also watch for generally suspicious items. For instance, in Afghanistan many roadside "gas station" shanties can be found. These places sell fuel oil, fertilizer, and antifreeze to law-abiding citizens; however, these items can be used to make IEDs. Alone, they do not indicate danger, but when stocked in unusual quantities or when they are the sole inventory of a store, then they may signal danger.

Latin Kings street gang graffiti

Atmospherics to Watch: When entering any area, personnel should ask themselves the following atmospherics questions:

- Does it look or feel or smell a certain way? Is it better or worse than a few minutes or days ago?
- Do people's clothing fit the scenario? Do the emotions fit the scene? Do the vehicles or rickshaws fit the area?
- Are the people in this area aware of the presence of surveillance, insurgents, VBIEDs, snipers, or other dangers?
- Is visible rubble, trash, stickers, or graffiti present? Is the graffiti sophisticated or unsophisticated? Is the graffiti public or private?
- Is visible security present, such as motion lights, window bars, video cameras, or concertina wire?

Discussion Questions:

1. Have you experienced an atmospheric shift? Describe it.
2. Why is a private behavior (such as a private phone conversation) in a public location (such as a crowded market) an atmospheric cue?

5006: Iconography and Symbolism

Intermediate
Lecture and discussion
About 180 minutes

Terminal Objective: This module helps trainees become more aware of the significance of icons and symbols. Upon completion of this module, trainees will be able to explain the importance of icons and symbols, in general, and they will be able to list specific icons, symbols, and their meanings.

Resource DVD:

- 5006.1 – *Islam and Symbolism*, PDF
- 5006.2 – Icons, Video
- 5006.3 – "Hospital," Video
- 5006.4 – Icons Symbols, PPT

Instructor Activities:

1. Ask trainees to review *Islam and Symbolism*
2. Review the "Hospital" video
3. Discuss iconography and symbolism, and its importance
4. Discuss the images in the Icons and Symbols presentation

Synopsis of Module Topics:

Introduction: Most cultures, including criminal and insurgent sects, use icons, color, or other symbols to convey meaning. Personnel can gain an advantage merely by knowing to look for such symbols, and those who can interpret or "read" the symbols gain additional critical insights.

Icons: Pictures with identifiable meanings are called icons. Most cultures, religions, and organizations use icons. For instance, Islamic icons include:

- *Star and Crescent:* A common symbol of Islam, it represents the faith
- *Qur'an:* An image of the Qur'an, the holy book of Islam

The Mexican Mafia may use:

- *Black Hand:* The black hand symbol, copied from the Italian Mafia.

Also, universally understood icons exist:

- *Weapons:* The use of weapons, including swords or rifles, universally implies aggression and a willingness to fight. Such symbols are often associated with criminal or insurgent groups.

Color Significance: Most cultures attach some degree of symbolic meaning to different colors. For instance, in Western societies the color red is often associated with stop signals and green with go signals. The combination of red and green, however, may take on a different, religious, meaning and may remind viewers of Christmas. Like Western cultures, Islamic societies have shared concepts of color, and Islamic societies have a long history of using color for symbolic meanings.

- *Green:* Color of Islam (very special for Muslims)
- *Yellow:* Promise of heaven, piety, and forgiveness
- *Blue:* Promise of water (critical for people who live in a desert)
- *Black:* Color of martyrdom and murder
- *White:* Color of purity
- *Red:* Sacrifice (including people who are willing to sacrifice for jihad)
- *White on Black:* Radicals use white-on-black to imply "through martyrdom, we achieve purity." This is a dangerous color combination, and when personnel see it, they should consider it an anomaly.

Tattoos: Body art, tattoos, or "patches" may display significant meanings. Gangs often use tattoos to signify allegiance.

- *Southwest/Mexican Gangs:* Mexican gangs often wear the eagle from the Mexican flag. Specific gangs use additional identifiers. For instance, Mara Salvatrucha gang members wear "M.S." or "13" (because 'M' is the thirteenth letter of the alphabet). The Border Brothers gang, who frequently crosses the US–Mexico border, uses an Aztec warrior face or the number "22" (again, because 'B' is the second letter in the alphabet, thus, two-two means "BB").
- *White Supremacist Groups:* Supremacist groups often sport Celtic runes or symbols, such as the Celtic cross. They may also wear neo-Nazi icons, including the double lightning-bolt "SS" or a swastika. Supremacist organizations use letters or numbers, such as 88 (because 'H' is the eighth letter of the alphabet, "HH" stand for *Heil Hitler*), 311 (because 'K' is the eleventh letter of the alphabet and 'K' three times becomes "KKK"), or 23 (because 'W' is the twenty-third letter and means white supremacy).
- *Prison Tattoos:* Other tattoos imply the wearer served prison time; for instance, inmates (particularly white supremacists) may have a spider web tattooed on their arm or elbow. Of course, others may wear this tattoo, too, without realizing its meaning. Another common prison icon is a tear-drop, inked onto a person's face.

Flag of Pakistan

Previous Flag of Iraq

Flag of Afghanistan

Other Icons: Other visible markers, such as jewelry, clothing, or make-up, may signify allegiance to a particular group or ideology. For instance, Islamic men's headgear may suggest their tribe or station. A Shi'a wearing a black turban is likely well educated, while a Muslim wearing a green turban is proclaiming is he is an elder (that is, a *Hajji*) who has completed the *Hajj* (or pilgrimage to Mecca). Other outward signs, including clothing, jewelry, and even hand symbols may have symbolic meaning. For instance, holding up an index finger (the "one" sign) is the "Sign of Islam" in Middle Eastern countries; however, it may be used by insurgents to signal their commitment to do violence for Allah.

Discussion Questions:

1. Nearly all tattoos have some significance, although some tattoos have a private meaning known only to their wearer. Look around the classroom; what tattoos or other icons/symbols are the trainees wearing? What is the meaning of each?
2. Consider your area of operation; what icons or symbols are often seen in that region? What significance do they hold?

5007: Heuristics

Intermediate
Lecture and discussion
About 90 minutes

Terminal Objective: This module introduces trainees to the heuristics domain and helps them helps become more aware of their own heuristic biases. Upon completion of this module, trainees will be able to define "heuristics" and describe the danger of heuristics.

Module Vocabulary:

- Heuristics
- Tactical Shortcut

Resource DVD:

- 5007.1 – Heuristics, Video

If it walks like a duck, talks like a duck, and lays eggs like a duck, then it's a duck—unless it's a platypus.

Instructor Activities:

1. Describe the *heuristics* domain
2. Review the Heuristics video
3. Discuss the benefits and dangers of relying upon heuristics

Synopsis of Module Topics:

Introduction: Heuristics are the final domain of combat profiling. The heuristics domain is unique, and it uses all of the other domains.

Heuristics: *Heuristics* are rules-of-thumb or *tactical shortcuts.* The brain uses heuristics to find "close enough" matches. In other words, whenever someone describes a person, place, or activity by saying "he looks like a __________," "she reminds me of __________," or "it looks like a __________," that is a heuristic match.

Heuristics Example: A Marine on post looks down a street and sees a car parked on the side of the road. The driver exits the vehicle and pulls out a piece of paper. He unfolds the paper across the hood of the car and then orients his body to the paper; he then waves down a pedestrian, who points to the paper. In this case, it is reasonable to assume the men are looking at a map. This logical leap is a heuristic. The Marine does not need to personally observe the piece of paper to intuit that it is a map, based upon the people's actions and the Marine's own knowledge of human behavior and maps.

Danger of Heuristics: A heuristic match occurs whenever some stimulus triggers a mental file-folder; that is, whenever personnel think, "Oh, I've seen that before; I know what's going on." The danger is that criminals and terrorists will try to use heuristics to their advantage; they use heuristics to try to hide in plain sight. For instance, an insurgent may place an IED in a car along the side of the road, and then make that car appear to be broken down (e.g., raising the hood). The IED trigger man may be standing nearby, looking annoyed and appearing to talk on his cell phone. This scene appears to be an average broken down vehicle, so personnel may ignore it.

Discussion Questions:

1. What are other names you could use for the term "heuristics"?
2. How would you explain the dangers of relying too strongly on heuristics to your fellow personnel?
3. How are the concepts of *heuristics* and *urban masking* related? How do criminals use heuristics to hide in plain sight? Can you give any examples of this from your own operational environment?

5008: Reading Human Terrain (Activity)

Intermediate
In class exercise
About 180 minutes

Terminal Objective: This module gives trainees the opportunity to practice the six domains. Upon completion of this module, trainees will be able to deeply describe the six domains, apply their knowledge of the domains, and profile in controlled environments.

Resource DVD:

- 5008.1 – Reading the Human Terrain, PPT

Instructor Activities:

1. Use the presentation provided on the resource DVD
2. As a class, profile each of the images
3. Discuss each domain and point out relevant cues in each photo
4. Discuss the operational narrative of each image
5. Ensure trainees "prove it;" make them explain how they arrived at their operational narratives

Synopsis of Module Topics:

Photo Interpretation Exercise: Instructors should lead the trainees through a group discussion of each of the photos. Make sure that trainees use and understand the CODIAC vocabulary. For each photo, ask trainees to point out:

- Biometric cues, including identifying individuals' emotional states
- Kinesic cues, also identifying individuals' emotional states
- Proxemic cues, including identifying the leader
- Geographic cues, including habitual areas and anchor points
- Atmospheric cues, including public and private "signs"

Ask trainees:

- What icons or symbols do you see? What do they mean?
- Is anyone using urban masking?
- Is anyone dressed differently than the others? How so?
- Is anyone making overt gestures or symbols?
- Is anyone displaying signs of wealth, power, or leadership?
- Is everyone in the image comfortable?
- Is anyone nervous, angry, or annoyed?
- If there are any personnel in the image, what is their SA?
- Do you see any anomalies that need to be investigated?
- Would you kill, capture, or contact anyone in this image?

Mental Simulation: Additionally, encourage trainees to develop an operational narrative for the activities they think are occurring in the scene. What might be happening? Most importantly, make trainees "prove" their conjectures. If they think that someone looks uncomfortable in an image, make them explain what heuristics they used to make that determination.

Unit 6: Reading the Physical Terrain

6

Tracking is the art of reading the physical environment to identify evidence a person has left behind. Being able to read the physical terrain enables personnel to develop a better intelligence picture in regard to an enemy's size, activities, location, composition, equipment, and intent. This unit will introduce trainees to the art of combat tracking and how it can support CODIAC skills. After completing Unit 6, trainees will know practical ways to analyze the physical terrain and use their tracking skills to support a range of missions types.

Suggested Prerequisites:

- CODIAC Units 1–5

Terminal Learning Objective:

- At the completion of this unit, trainees will have intermediate knowledge of tracking techniques and basic knowledge of *tactical* tracking. This should enable them to use their tracking within their standard missions and/or under the guidance of more experienced trackers.

Enabling Learning Objectives:

- Explain how, when, and why a combat tracking team might be employed
- Describe the common combat tracking terminology
- Describe and demonstrate how to read spoor
- Describe and demonstrate interpretation of a scene by analyzing footprints
- Describe the characteristics of human pace
- Demonstrate how to estimate the number of people in the quarry based upon their spoor
- Describe methods of assessing the age of spoor

Estimated Time Allotted for Instruction: About 21 hours

Speciality Support Requirements:

- Spoor Pit (preferably located near classroom) for demonstrations and a rake to clear the pit
- Ranges or other large, open locations that will support follow-up practical exercises
- For Units 6 and 8, training participants should acquire copies of the book *Tactical Tracking Operations* by David Scott-Donelan. This publication is an important resource for learning combat tracking.

6001: Introduction to Combat Tracking

Introductory
Lecture and discussion
About 30 minutes

Terminal Objective: This module introduces trainees to the purpose and general practices of combat tracking. Upon completion of this module, trainees will be able to describe, in general terms, the typical goals, mission, and employment of modern day combat tracking teams.

Module Vocabulary:

- Combat Tracking
- Man Tracking

Resource DVD:

- 6001.1 – Introduction to Combat Tracking, Video
- 6001.2 – Combat Tracking Skills, Video

Other Suggested Materials:

- *Tactical Tracking Operations*
  — Chapter 1
  — Chapter 2

Instructor Activities:

1. Begin by asking trainees what they think "trackers" do
2. Describe the purpose and goals of a combat tracking team
3. Review the two videos provided on the Resource DVD
4. Present real-world anecdotes describing combat tracking

Synopsis of Module Topics:

Introduction: What does "tracking" mean? Do only hunters use tracking? What is modern tracking used for? Did you know that we have trackers in Afghanistan and Iraq, and on the US–Mexico border, who are mission critical and performing with great distinction?

Combat Tracking: Effective combat trackers are significant assets for offensive operations, intelligence collection, and clandestine movements in hostile areas. They gather information by analyzing the evidence left by individuals on the physical environment, and this information allows personnel to develop a better intelligence picture in regard to an enemy's size, activities, location, composition, equipment, and intent.

Combat Tracking Purpose and Goals: The goals, missions, and employment of a combat tracking team will vary based upon the specific operation. Tracking teams are most often employed to (1) follow the tracks of armed aggressors, (2) discover tracks by the use of patrolling and reconnaissance techniques, (3) place pressure upon aggressors during a follow-up to drive them into ambushes and prepared positions, (4) locate and interpret tracks left by enemy activities, (5) ascertain the direction of flight of insurgents so as to better concentrate blocking forces, (6) recognize and calculate strengths of aggressor patrols and formations, (7) maintain contact with a fleeing enemy, and (8) help recon and sniper teams move in and out of denied or hostile territory without leaving footprints or other evidence (by using anti-tracking skills and techniques).

Discussion Questions:

1. Consider the eight main uses of a combat tracking team; what skills does a team need to most effectively accomplish each task?
2. How are combat tracking skills relevant for CODIAC? How might combat tracking inform decision-making? Situational awareness? Analyzing the human terrain?
3. What makes *combat* tracking different from other forms of tracking, such as searching for a lost child or stranded hiker? What unique challenges does a combat tracker need to consider?

TRY THIS!

Perform an Internet search using the terms "Combat Tracking" or "Man Tracking." What sort of articles do you find? Did you find any recent news articles about combat tracking?

6002: Combat Tracking Terminology

Instructor Activities:

1. Discuss the advantages of unambiguous communication
2. Describe the reasons for specific technical terms
3. Review the Terms video

Introductory
Lecture and discussion
About 60 minutes

Terminal Objective: This module introduces specific combat tracking terminology. Upon completion of the module, trainees will be able to describe the key combat tracking vocabulary.

Module Vocabulary:

- Active Track
- Contamination
- Follow-up
- Ghost Spoor
- Initial Commencement Point
- LiNDATA
- Lost Spoor Procedures
- Natural State
- Passive Track
- Quarry
- Spoor
- Spoor Cutting
- Spoor Separation Point
- Time and Distance Gap
- Track Line
- Tracker Support Team
- Tracking Team

Resource DVD:

- 6002.1 – Terms, Video

Other Suggested Materials:

- *Tactical Tracking Operations*
  — Chapter 4
  — Chapter 2

Synopsis of Module Topics:

Introduction: Personnel should have a complete understanding of the unique tracking terminology to ensure communications are free of ambiguity. Each basic vocabulary term can be discussed based upon which main combat-tracking task it supports:

- Establish the track
- Interpret and follow the track
- Gain intelligence to develop a track picture (LiNDATA)
- Operate in hostile territory without leaving visible track (spoor)
- Encounter and react to enemy contact

Gaining Intelligence Information: The success of the tracking mission is often dependent on the amount of relevant information obtained by the tracking team. Critical ("must have") information regarding the quarry can be conveyed as a LiNDATA SitRep (Situation Report):

- *Location*: The Initial Commencement Point (e.g., a GPS plot)
- *Number*: The quantity of people to be tracked
- *Direction*: The initial direction of flight (e.g., Cardinal direction)
- *Age*: The age of the tracks (i.e., the time/distance gap)
- *Type*: The type of footwear patterns
- *Additional*: Any additional relevant information ("should have") that ought to be collected, such as distinguishing features about the quarry, physical descriptions, and eye-witness reports. Also, if possible, the tracking team should learn about the history of an area (e.g., families/tribes, local religious sects) and the typical tactics used by the enemy (e.g., use of anti/counter-tracking, enemy habits, and preferences).

Discussion Questions:

1. Describe different ways in which a site may become contaminated. How might your fellow personnel unwittingly contaminate a site?
2. What is the difference between active and passive tracking?
3. How can tracking support tactical and strategic operations? How might the use of tracking benefit the missions your organization conducts in your current operational environment?

6003: Reading Spoor and Sign

Introductory
Lecture and discussion
About 90 minutes

Terminal Objective: This module introduces the process identifying spoor. Upon completion of this module, trainees will be able to describe the indicators left by a person in a spoor pit or at a practice range, discuss the principles of time/shadow effect, and position themselves to maximize the visibility of the spoor based upon the angle of the sun.

Module Vocabulary:

- Aerial Spoor
- Color Change
- Conclusive Evidence
- Desert Shine
- Disturbance
- Flattening
- Ground Spoor
- Litter
- Regularity
- Sign
- Substantiating Evidence
- Time/shadow effect
- Transference

Resource DVD:

- 6003.1 – Indicators, Video

Other Suggested Materials:

- *Tactical Tracking Operations* — Chapter 3, pages 29–37

Instructor Activities:

1. Discuss the need to recognize action indicators
2. Describe and discuss the visible indicators
3. Describe and discuss the non-visible indicators
4. Describe how light and shadow affect viewing ground spoor
5. Describe undesirable effects of tracking too close to oneself
6. Review the Indicators video

Synopsis of Module Topics:

Tracking Indicators: When a person passes over a piece of ground, his/her passage is marked by three types of indicators: ground spoor, aerial spoor, and sign. Tracker should use some combination of these indicators to track the quarry. Depending on the type of terrain and the skill/number of quarry there will be great variations in the type of indicators left behind.

Ground Spoor: Ground spoor consists of any mark or indication left on the ground by footwear, body parts, or equipment. Ground spoor is defined by any of the five following characteristics:

- *Regularity:* Man-made patterns, such as prints with a uniform tread
- *Flattening:* Marks left on the ground by the weight of a foot, etc.
- *Transference:* Dirt/vegetation carried from its natural location
- *Color Change:* Subtle color changes caused by ground disturbance
- *Disturbance*: Ground that has been moved from its natural state

Similarly, tracks are identified by their:

- *Outline:* These are outer limits, or edges of the track
- *Shape:* The recognizable form of something man-made
- *Color:* The disturbance of the natural area will cause color change
- *Texture*: The natural smoothness/roughness is affected by the quarry
- *Shine*: Disturbing the ground causes light to reflect differently
- *Rhythm:* Nature has its own rhythm, but usually without regularity; something in nature that is spaced at regular intervals will stand out as an anomaly from the natural state of the environment

Aerial Spoor: Aerial spoor consists of damage or disturbance to the surrounding vegetation, from foot to head height, such as crushed or bent grass, crushed or bruised leaves, or twisted vegetation or vines. Aerial spoor is a form of *substantiating evidence*; it should always be confirmed by identifying nearby *conclusive evidence*.

Sign: Sign encompasses the whole group of indicators that are not part of ground or aerial spoor, such as broken cobwebs, disturbed insects nests, or water splashes and deposits.

- *Litter*: Accidently dropped, discarded, or hidden artifacts
- *Body Waste:* Any naturally occurring matter produced by the human body, including urine, feces, and oral ejecta (anything spat out)
- *Blood Spoor:* Blood spoor consists of any blood dropped or splashed onto the ground from a wound
- *IED Indicators:* Clues indicating that a booby trap or anti-personnel mine has been placed

Non-Visible Indicators: Non-visible indicators include noises, smells, and other sensory activators.

- *Smells and Odors:* Sweat, bug spray, or rifle oil can be smelled from several yards away, and cigarette smoke and cooking odors can be smelled for up to a thousand yards away.
- *Unnatural Noises for the Specific Environment:* Noise (or lack of noise) may indicate the presence of the quarry. Talking, whistling, metallic sounds, and chopping can be heard at great distances and can give early warning to the tracker that the quarry is near. On the other hand, insects ceasing to chirp or buzz, or sudden silences in a rural environment, may also indicate the presence of the quarry.

Light and Shadow: Light and shadow determine the visibility of spoor on the ground. The lower the light source is in relation to the ground, the more shadow will be created by the imprint of the spoor. This greatly enhances one's ability to spot spoor. The best time to spot tracks is in the early morning or late afternoon light, when the sun is at a 45° angle to the tracks. Spoor is most difficult to follow when the sun is at its peak, shining straight down on the tracks and throwing little or no shadow. When tracking, place the tracks between the tracker and the sun to maximize the tracker's ability to see the shadow formed by the spoor.

Discussion Questions:

1. What is the difference between conclusive and substantiating evidence? Give an example of each.
2. What can be done by the tracker to enhance the light/shadow effect on the tracks when the ambient lighting is not optimum?
3. Should tracking ever be attempted at night? If so, how can the spoor and sign be seen? Should you use high-powered flashlights?

REMEMBER!

Best tracking times:

- First light to 1030
- 1530 to sunset

Worst tracking times:

- The hour approaching midday and the hour after midday when the sun is high in the sky

"TRACKS ARE CLUES, the most clues a perpetrator will leave behind. One every thirty inches or so and as conclusive as fingerprints."
—*Sherlock Holmes*

6004: Dynamics of a Footprint

Introductory
Lecture and discussion
About 120 minutes

Terminal Objective: This module discusses the intelligence that can be gathered by interpreting individual footprints. Upon completion, trainees will be able to describe the features of a footprint.

Module Vocabulary:

- Balance
- Dwell Time
- Foot Roll
- Lugs/Grippers
- Heel Strike
- Pace
- Pitch Angle
- Pressure
- Pre-Terminal Point
- Primary Impact Point
- Rhythm
- Straddle
- Stride
- Terminal Point
- Toe Dig

Resource DVD:

- 6004.1 – Micro-Tracking Introduction, Video
- 6004.2 – Micro-Tracking Demonstration, Video
- 6004.3 – Footprint, Video

Other Suggested Materials:

- *Tactical Tracking Operations* — Chapter 3, pages 38–40

Instructor Activities:

1. Define the different components used in footprint interpretation
2. Demonstrate the interpretation of drawn/photographed prints
3. Review the Micro-Tracking Introduction video
4. Review and discuss the Micro-Tracking Demonstration video

Synopsis of Module Topics:

Every Step a Word... A tracker tries to build a story by looking at a set of tracks. By developing this story the tracker can understand the who, how, and when of what happened. The first step in building the story consists of reading an individual print.

Reading Footprints: Each footprint displays a set of indicators, and from these a skilled tracker can interpret the actions of the quarry. These footprint features include: the *primary impact point* (PIP), *foot roll, heel strike, pre-terminal point* (PTP), and *terminal point.*

Footwear Types: Footprint patterns are cataloged as western, heels, flats, cleats/lugs, and barefoot. Each set of prints should be given a nickname, to facilitate discussion and make continued identification easier.

Characteristics of Human Pace: It is possible to interpret with some accuracy what activity has taken place, based upon interpreting the elements of a human pace. These are

- *Stride:* Distance from one heel to the other heel
- *Pitch Angle:* Angle the foot "pitches out" from the line of travel.
- *Straddle:* Distance between the two feet at the closest point
- *Pressure:* Weight of the body, through the foot, onto the ground
- *Dwell Time:* Amount of time a foot is on the same ground location
- *Rhythm:* The measured regularity of footprints
- *Balance:* Maintaining center of gravity evenly, with minimal sway

Average Stride: An adult's average stride spans about 30 inches.

Estimating Height: Roughly, the height of a person can be determined by measuring the length of his/her footprint. Measure the footprint in inches, then divide this value by 2. This value is roughly the height of the quarry in feet.

Discussion Question:

1. What can a tracker learn about his quarry from a few footprints? What types of missions could benefit from such intelligence?

6005: Micro-Tracking (Activity)

Introductory
Spoor Pit Exercise
About 120 minutes

Terminal Objective: This module allows trainees to practice micro-tracking in a controlled setting. Upon completion, trainees will be able to demonstrate basic footprint/pace interpretation.

Module Vocabulary:

- Micro-tracking

Resource DVD:

- 6005.1 – Micro-Tracking Spoor Pit Exercise, Video
- 6005.2 – Footprint Analysis Demonstration

Instructor Activities:

1. Describe the concept of micro-tracking
2. Review the two videos provided on the Resource DVD
3. Go to the spoor pit; demonstrate micro-tracking techniques
4. Break trainees into teams and ask them to complete the exercise

Synopsis of Module Topics:

Micro-Tracking: Micro-tracking describes the process of identifying every spoor in sequential order. Micro-tracking is commonly referred to as the "step-by-step" method, and is typically used for search and rescue missions and specific evidence gathering. This method is very effective, but time consuming and tiring; it works only when closing the time-distance gap is not mission essential.

Sun Angle Demonstration: It is best to conduct spoor pit demonstrates in the morning or late afternoon, when the sun is at an angle to the ground. Ask trainees to walk around the spoor pit and determine from which side the tracks are most visible. Point out that the tracks are most visible when they place the tracks between themselves and the sun.

Micro-Tracking Demonstration: Demonstrate each of the dynamics of a footprint in detail. Point out each of the features of a footprint and of a few paces. Measure the length of a stride, and measure the length of a print and then calculate height from it. Ask trainees—particularly those of varied heights and weights—to also walk across the spoor pit. Measure their strides and footprint. Ask the trainees to examine their own, and fellow trainees', prints. What features can they identify. What nicknames would they use to describe the prints?

Left: David Scott-Donelan measures a footprint during a micro-tracking spoor pit demonstration

6006: Interpreting Spoor and Sign

Introductory
Lecture and discussion
About 90 minutes

Terminal Objective: This module introduces trainees to the action indicators they will need to recognize in order to interpret the actions of a quarry based upon his/her spoor. Upon completion of this module, the trainee will be able describe spoor interpretation processes, and he/she will be prepared to participate in a micro-tracking practical application.

Module Vocabulary:

- Action Indicators

Resource DVD:

- 6006.1 – Action Indicators, Part 1, Video
- 6006.2 – Action Indicators Part 2, Video

Instructor Activities:

1. Discuss the types of data that can be interpreted from spoor
2. Discuss action indicators and their importance
3. Review the two videos provided on the Resource DVD

Synopsis of Module Topics:

Introduction: Footprint patterns, IED materials, trash, and other spoor and sign—all of these tell a piece of the story. Skilled trackers can analyze the spoor and sign to interpret, with some accuracy, what activities took place at a given location. Trackers can interpret a quarry's...

- Speed
- Size/weight
- Gender
- Age and activity level
- Equipment carried
- Time of movement (i.e., moving by day or night)
- Weaponry
- Situational awareness
- Physical state (e.g., tired)
- Lower body disabilities
- Mental state
- Physical activities (e.g., pausing to look through optics)

Determining Speed: By determining the speed of the quarry, the tracker can also determine where the quarry will be at a certain time and can relay this information to blocking or intercepting forces. *Sprinting* quarry will leave tracks primarily made by the balls of his/her feet. They will be widely spaced and contain little, if any heel, impressions. *Fast moving* quarry will leave widely spaced, deeply imprinted tracks. There will be a more distinct heel impression and normally a scrape mark near the toe as it leaves the ground. Quarry moving *slowly and deliberately* will leave tracks that are evenly spaced and of uniform depth. The primary impact point and terminal impact point will be almost identical. Finally, the tracks of quarry moving *very slowly* will be close together and evenly spaced.

Action Indicator: Whenever the tracks change, that is an action indicator. Action indicators suggest that the quarry is engaging in a new activity, and any such change in behavior is worth considering, assessing to determine whether it has tactical relevance, and if relevant, communicating to the command element.

Discussion Questions:

1. Why should a tracker try to interpret the spoor and sign? Why is interpretation better than simply following the tracks?
2. Is there a time when you should forego interpretation? Why?

6007: Micro-Tracking 2 (Activity)

Instructor Activities:

1. Go to the spoor pit
2. Demonstrate spoor interpretation techniques
3. Break trainees into teams and ask them to complete the exercise

Intermediate
Spoor Pit Exercise
About 180 minutes

Terminal Objective: This module gives trainees the opportunity to learn and practice their spoor and sign interpretation skills. Upon completion of this module, trainees will have a working ability to analyze ground spoor.

Synopsis of Module Topics:

Micro-Tracking Demonstration: This demonstration should emphasize the spoor and sign interpretation skills that were presented in the previous section. First, the instructor should help trainees identify action indicators or other relevant cues in the spoor pit. Have trainees move across the spoor pit at various speeds, and then note how the footprints differ. If weapons (or simulated weapons) are available, take note of what ground spoor they leave. Demonstrate how the spoor and sign interpretation skills can be applied to various realistic scenarios.

Scenario Ideas: Practice scenarios are limited only by your imagination. Below are a few ideas, but this list is certainly not exhaustive. Also, be encouraged to bring gear or other "props" to view the spoor they leave in the pit; these may help you identify additional practice scenarios.

- Two individuals walk toward one another; they stop and exchange something, and then turn around and walk back the way they came.
- Someone is carefully creeping; he/she stops for a moment, as if listening, plants something (e.g., an IED), and then stalks away.
- Two people are walking together and one of them is carrying a heavy load (e.g., another trainee on his/her back).
- A team is moving together. They stop, and then the point-person takes a prone position, pulling out his/her optics to view the horizon. After that, the team continues on.
- A person is injured and disoriented. He/she favors one of his ankles.
- A person is moving at night, and he/she comes suddenly upon a tree. How does this track line differ from the same path during the day?

Micro-Tracking Exercise: After the instructor has demonstrated spoor and sign interpretation, divide the trainees into five-person teams. Assign two teams to a spoor pit area. Ask the first team to act out a scenario while the other team turns around and closes their eyes. Then the first team turns around and, as a group, attempts to determine what activities took place by interpreting the ground spoor. As trainees complete this activity, observe their progress. You may need to recommend scenario ideas to the spoor-laying teams.

6008: Counting the Quarry

Introductory
Lecture and discussion
About 60 minutes

Terminal Objective: This module introduces trainees to methods for determining the number of quarry being tracked. Upon completion of this module, trainees will be able to describe three methods for determining the number of quarry.

Module Vocabulary:
- Average Pace Method
- Comparison Method
- Direct Count Method
- Key Print

Resource DVD:
- 6008.1 – Counting, Video

Other Suggested Materials:
- *Tactical Tracking Operations* — Chapter 4, pages 56–57

Instructor Activities:

1. Discuss the importance of assessing the number of quarry
2. Describe the *direct count method*
3. Describe the *average pace method*
4. Describe the *comparison method*
5. Review the Counting video

Synopsis of Module Topics:

Importance of Counting the Quarry: Determining the number of quarry is critical for the tracking team. The life of each team member depends upon knowing the threats he/she may face, and team tactics will depend on the number individuals being tracked.

Choosing a Count Method: To determine the number of quarry being followed the tracker should first estimate the approximate number of quarry; this will inform which, more precise, counting method can be used. If the team appears to be tracking fewer than 6 people, then the tracker should use the *direct count method*, and the tracker should use the *average pace method* if the quarry appears to number between 7–15. If the quarry consists of more than 15 individuals, then the tracker should use the *comparison method*.

- *Direct Count Method:* As the name implies, *the direct count method* involves counting each quarry's individual footprints. Draw a line behind the heel of an easily recognizable print. This print becomes the *key print*. Draw another line behind the opposite foot of the *key print* (e.g., if the *key print* is a left foot, then the opposite foot is the right). Count each print between the lines (including the *key print*), this indicates the number of quarry.
- *Average Pace Method:* To use the *average pace method*, first determine a *key print* and then draw a line behind it. Draw another line at the heel of the next opposite foot, therefore measuring two stride lengths. Count all of the prints between the lines and divide by two to determine the number of quarry.
- *Comparison Method:* For the *comparison method*, the team walks alongside an unknown number of tracks in multiple passes (while counting each pass) until the known and unknown tracks have a similar appearance.

Discussion Question:

1. Why is it important to know how many quarry are being followed? How might this intelligence be useful to the tracking team?

6009: Micro-Tracking 3 (Activity)

Introductory
Spoor Pit Exercise
About 180 minutes

Terminal Objective: This module gives trainees the opportunity to learn and practice their quarry counting skills. Upon completion of this module, trainees will have a working ability to determine the number of quarry based upon their ground spoor.

Instructor Activities:

1. Demonstrate spoor interpretation techniques
2. Break trainees into teams and ask them to complete the exercise

Synopsis of Module Topics:

Counting Demonstration: This demonstration should emphasize the spoor counting techniques discussed in the previous section. The instructor should enlist the help of different numbers of trainees to lay the spoor, and then the instructor should lead the trainees through the counting process. Instructors are encouraged to bring track-sticks, yardsticks, string, or other tools to this demonstration to help mark-off strides for the *average pace* and *direct count methods*.

Counting Exercise: Following this, the instructor may break the trainees into two groups, depending upon the number of trainees. One group helps lay spoor, while the other group turns their backs to the pit. After each line of spoor is laid, the second group of trainees should attempt to count the number of personnel who left the track. As the class becomes competent with the basic quarry counting techniques, instructors are encouraged to begin introducing additional action indicators (such as those practiced during Module 2007). Ask certain trainees to carry a weight or ask several trainees to lay spoor in the same direction but act as if they are part of two separate groups, not traveling together. Again, be creative with the ideas used in these scenarios.

Left: Border Patrol agents lay a track line in a spoor pit during a quarry-counting exercise.

6010: Assessing the Age of Spoor and Sign

Introductory
Lecture and discussion
About 90 minutes

Terminal Objective: This module introduces trainees to assessing the amount of time that has elapsed since the quarry made the spoor. Upon completion of the module trainees will be able to describe the factors affecting the spoor over time, how the spoor reacts to those factors, and how to estimate the age of spoor based upon this knowledge.

Other Suggested Materials:

- *Tactical Tracking Operations* — Chapter 3, pages 40–46

Instructor Activities:

1. Discuss the importance of assessing the age of the spoor
2. Discuss the influencing factors on the age of observable spoor
3. Discuss the effect of age and time on spoor

Synopsis of Module Topics:

Introduction: In order to the close the time/distance gap and gather accurate intelligence, personnel must be able to estimate the age of the spoor. This allows one to accurately judge how much time has elapsed since the spoor was created. The age of the spoor is typically stated in two-hour increments. As a general rule, it is better to overstate the age of the spoor rather than to understate it. Also, personnel must be aware that the age of the spoor can change suddenly; for example, if the quarry started out six hours ahead, but stops for a four-hour rest, then the spoor will go from being six hours old to two hours old.

Aging Spoor: The best way to learn how to assess the age of the spoor is by constant practice, but even experienced trackers can be wrong in their assessment. Personnel should use as many factors as possible in order to judge the age or spoor. Relying on a single factor can often lead to misleading results.

Weather: Once the spoor is left, it will begin to erode. Environmental factors determine how quickly that erosion takes place. It is important for trackers to remember when different weather effects occurred. For example, a light rain may round the edges of the print. By remembering when the last rain fell, the tracker can place the print into a time frame.

Animals: Animal activity can also be useful in judging the age of spoor. Most wild animals will make their way to water sources in the morning and evening, so spoor age can be determined by whether or not the spoor is overtop or underneath animal tracks.

Plants: Vegetation can play a key role in judging the age of spoor. Most green vegetation will start to dry out and brown quickly when broken or damaged. Trampled vegetation that is still moist and green indicates the spoor is very fresh.

Aging Stands: One of the most accurate techniques used to learn to assess the age of spoor and sign, is to create "aging stands." Under controlled circumstances, personnel can observe sets of spoor and sign, footprints, damaged vegetation, fire pits, and food products for observation over an extended period of time to see how they alter in appearance over time.

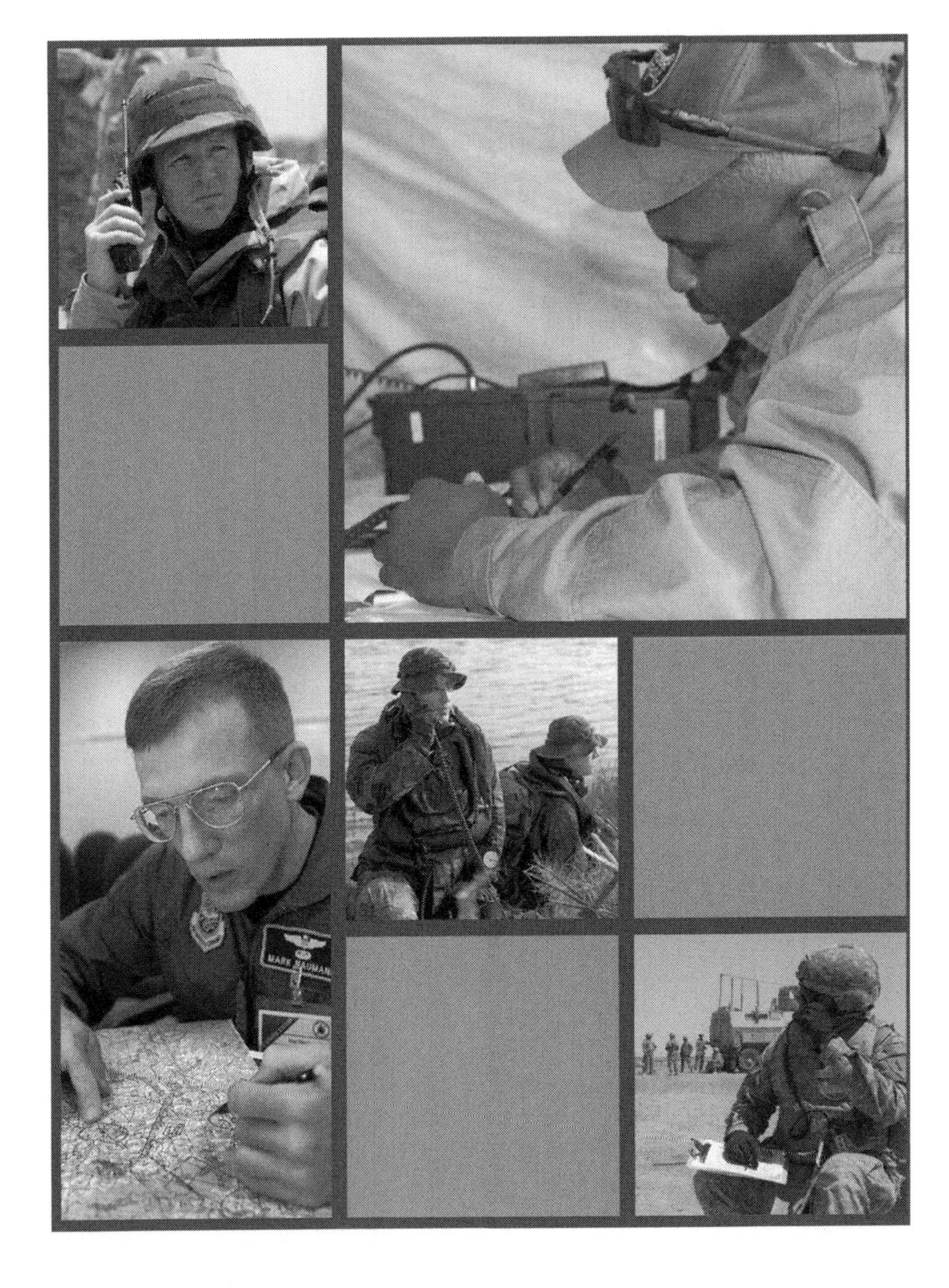

Unit 7: Operational Intelligence Cycle

7

Military and law enforcement personnel never work in isolation. Their operations are conducted as part of larger and wider missions. Being able to observe, understand, and then communicate actionable intelligence based upon the human and physical terrain enables units to develop much more detailed intelligence pictures, and, ideally, react to threats before they are triggered. This unit will introduce trainees to the wider operational picture, in terms of support personnel and command elements; trainees will learn strategies for effectively communicating CODIAC information to/from these elements. After completing Unit 7, trainees will better understand how their use of CODIAC skills fits into the larger operational environment, and they will be able to demonstrate effective communication of CODIAC intel.

Suggested Prerequisites:

- CODIAC Units 1–6

Terminal Learning Objective:

- At the completion of this unit, trainees will have detailed knowledge of the operational employment of CODIAC skills, and they will be able to describe tactically relevant observations and interpretations to other distributed teams, tactical operations centers, and command elements.

Enabling Learning Objectives:

- List and define the steps of the intelligence cycle
- Explain how tactical teams contribute to the intelligence cycle
- Describe several ways that tactical teams can improve their contributions to the intelligence cycle
- Demonstrate effective communication of CODIAC intel to/from a tactical operations center

Estimated Time Allotted for Instruction: About 5 hours

Speciality Support Requirements:

- No special facilities required

7001: Intelligence Cycle

Introductory
Lecture and discussion
About 60 minutes

Terminal Objective: This module introduces the intelligence cycle, which describes the process of transforming collected information into usable intelligence. Upon completion of this module, trainees will be able to describe the five phases of the intelligence cycle.

Module Vocabulary:
- Collection
- Processing and Exploitation
- Production
- Dissemination
- Planning and Direction

Resource DVD:
- 7001.1 – *Joint and National Intelligence*, PDF

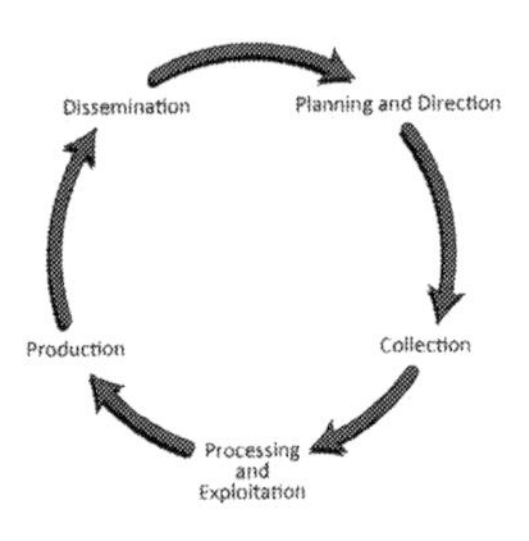

Above: Intelligence cycle diagram

Instructor Activities:

1. Discuss the five phases of the intelligence cycle
2. Discuss how the intelligence cycle works in a circular manner
3. Discuss how small units contribute to the intelligence cycle
4. Discuss the importance of providing information into the cycle

Synopsis of Module Topics:

Introduction: Small, dismounted teams make up an integral part of the intelligence cycle. Any information that a tactical team collects is potentially valuable intelligence, especially when combined with information collected by other sources. However, to be maximally effective, the intelligence collectors must have a working understanding of the intelligence cycle and how they fit into the intelligence collection process.

Collection: Often, a small tactical unit is the first to collect information on the terrain, area of operations, hostile forces, or local populace. During the *collection phase* of the intelligence cycle, those personnel who collect information must accurately identify relevant data and communicate it (correctly and completely) to the command element. With CODIAC training, trainees are now able to communicate additional, nuanced data to their command elements. Such information should be actively included in trainees' SitReps, from now on.

Processing and Exploitation: As soon as information is communicated to a command element it enters the *processing and exploitation phase* of the intelligence cycle, where it begins transformation from "information" into "intelligence." *Information* describes the raw observations collected in the field (e.g., an AK-47 taken from a hostile insurgent), whereas *intelligence* is the product that can be gleaned from interpretation and application of information (e.g., several other AK-47s have been captured, and when examined, it was determined they all came from a shipment that was sold by Country X).

The processing and exploitation phase is conducted at all levels, from the company to the battalion, and potentially even higher. For example, if a tactical team in Afghanistan captures a member of the Taliban with several photographs of an American nuclear power plant, that information would travel all the way to the Department of Homeland Security for processing and exploitation. How the information was labeled, correlated, and prioritized by the tactical team (during the collection phase) will affect the success of the processing and exploitation phase.

Production: After initial information is processed it enters the *production phase*, where the significance of the intelligence is established and its implications determined. In this phase, intelligence is combined with data from additional sources, including other small tactical teams. From the combined intelligence, an operational picture begins to develop, and command elements (who cannot directly see to the field) gain greater understanding of the battlespace.

Dissemination: The combined intelligence picture then enters the *dissemination phase* where command elements communicate it, as needed. For instance, intelligence may flow back down to tactical levels, and small units may receive the intelligence gleaned through the information cycle. The tactical team should be proactive during this phase, requesting the information it needs for each operation. Just as the commander of a tactical team is responsible for procuring the required weapons, equipment, and rations for an operation, he/she is equally responsible for procuring the necessary intelligence for an area of operations. Relevant intelligence may include satellite photos of the battlespace, weather reports, and intelligence on the local population.

Planning and Direction: Intelligence in the *planning and direction phase* may be high-level and managed by upper-level government agencies, or it may be low-level and disseminated back to the small-unit tactical teams. Using the example above, photos of the nuclear power plant would be sent to DHS and dealt with at a high level, and tactical teams would not participate in the planning and direction of that particular intelligence picture. However, in the other example, the intelligence that a shipment of AK-47s came from Country X might be disseminated to tactical teams, enabling them to adjust their tactics to prepare to encounter additional firepower from that country.

"SUCCESS IN DEVELOPING information superiority depends upon integrating information from a range of sensors, platforms, commands, and centers to produce all source intelligence. This intelligence must be part of a portrayal of the battlespace characterized by accurate assessments and visual depiction of friendly and enemy operations which makes the battlespace considerably more transparent for a United States commander than for the adversary and forms the basis for superior decision making."

—RADM L. E. Jacoby, USN
Quoted in Joint and National Intelligence (page III-55)

Discussion Questions:

1. Small-unit leaders should proactively request intelligence before operations. What does this mean? Give an example of how a small-unit leader can request intelligence.
2. How does participation in a broad intelligence cycle benefit a unit more than if they only gathered/analyzed their own information?
3. In *Intelligence Analysis: A Target-Centric Approach*, Robert M. Clark suggests the intelligence cycle should be conceptualized as a network where personnel at each node collaborate—rather than following linear communication channels. Do you agree? Why or why not?

7002: Intelligence Collection

Intermediate
Lecture and discussion
About 90 minutes

Terminal Objective: This module discusses intelligence collection activities from the perspective of CODIAC skills. Upon completion of this module, trainees will be able to describe the small unit, tactical information gathering process.

Module Vocabulary:
- Common Tactical Picture
- HUMINT
- Site Exploitation

Resource DVD:
- 7002.1 – Intelligence, Video
- 7002.2 – *DOD 3115.09*, PDF
- 7002.2 – *Intelligence Exploitation*, PDF

REMEMBER!

To be most effective, personnel must select, use, and maximize the use of available optics (from the naked-eye to advanced optical systems). Then, through attention to detail, each person can establish a baseline of an environment and detect the anomalies, which are critical pre-event indicators.

Instructor Activities:

1. Reiterate the intelligence collection processes already discussed
2. Discuss the intelligence collection steps
3. Review the Intelligence video
4. Through interactive discussion, help trainees integrate the CODIAC skills into intelligence collecting tactics, techniques, and procedures

Synopsis of Module Topics:

Introduction: In addition to honing individual observation skills, personnel and small units must acknowledge the value of shared observations. All military and law enforcement agents must strive to create a *common tactical picture*, driven by a common language, in order to articulate information and effectively share intelligence.

Step 1 – Establish a Baseline: Establish a common baseline of the environment. Discuss the baseline and create a common tactical picture within the team. Be alert for anomalies that rise-above or fall-below the baseline, and look for template matches of specific items or people of interest (such as High Value Targets or BOLOs).

- *Sketching and Photographing:* Taking photographs of the people and key areas of a region can be useful for comparison or later reference. Similarly, sketching an area can help personnel develop clearer baselines and identify subtle anomalies.
- *Denial:* Remember, each person's primitive brain will attempt to make order from chaos, often through denial; personnel must train themselves to *actively look* for important cues and pre-event indicators.
  - — Denial is a "tab" on every person's first mental file-folder
  - — Denial is the most dangerous phase of decision-making
  - — Denial occurs *every time* a "non-standard" observation is made

Step 2 – Seek Anomalies: Actively *hunt* for anomalies. Constantly measure everything against the team's baseline. Discuss what each team member sees; think aloud and discuss operational narratives. Remember, adversaries hide in plain sight, use urban masking, and attempt to exploit the seams and gaps of a battlespace.

- *Investigate Every Anomaly:* Every anomaly must be investigated; err on the side of action. In other words, do not lapse into a state of denial. Treat every anomaly as a potential danger indicator.
- *Site Exploitation:* Upon entering a sensitive site (such as a suspect's home) personnel must immediately collect, evaluate, prioritize, and disseminate the information of immediate/potential tactical value.

- *HUMINT:* Tactical teams should also consider collecting immediate and long-term human intelligence (*HUMINT*) from civilians in their operational area.

  Immediate HUMINT examples:
  - — Are any strangers in the village today?
  - — Did you see anything unusual yesterday or today?
  - — Have any trucks gone down this road today?

  Long-term HUMINT examples:
  - — How many people live here? How many men?
  - — Do you ever get truck traffic through here?
  - — Who controls the local area?

Step 3 – Decide: Once three anomalies (or one major anomaly) have been identified, a decision must be made: Kill, capture, or contact. Remember, each decision must be legal, moral, and ethical; it must follow the Rules of Engagement (ROE) and procedures for Escalation of Force (EOF).

Step 4 – Communicate: Communicate decisions within the team and to command elements. As appropriate, consider communicating important information to lateral teams. When communicating intelligence:

- Distinguish between raw information and personal analyses.
- Be prepared to articulate the decision and its rationale ("prove it").
- Use effective and efficient communication.
- Have a pre-established communication plan for the teams operating in an area; otherwise communications will become chaotic—especially if multiple teams observe the same activity at the same time.

Discussion Questions:

1. What are some ways that making sketches or taking photos might help you in your current operational tasking?
2. Think of the missions you most frequently conduct. How might better HUMINT help you attain mission success? How about the most critical operations, what HUMINT would be most valuable?
3. Think of your organization and your operational environment. What is an appropriate communication plan? Is it ever appropriate to communicate to lateral teams?
4. How much communication does CODIAC data require, do you think? Will you need to communicate more or less than you do already? What other ways might your communications change?

DID YOU KNOW?

Both the Army and Marine Corps are pursing efforts that emphasize the intelligence collection abilities of individual Soldiers and Marines. The Army's program is called *Every Solider a Sensor*, and the Marines' have dubbed their effort *Every Marine a Collector.*

KEY SKILL

Developing an internal sense of time in order to know when a situational judgment needs to be updated

Instructors should emphasize the importance of continuously updating one's baseline. Discuss how often personnel should actively reevaluate their baselines. Discuss the importance of having an internal sense of time, so that individuals know when to update their own operational pictures.

7003: Tactical Operations Center

Introductory
Lecture and discussion
About 60 minutes

Terminal Objective: This module introduces trainees to the concept of the Tactical Operations Center (TOC). Upon completion of this module, trainees will be able to describe the function of a TOC and the advantages of distributing intelligence and operations functions from the battalion to the company level.

Module Vocabulary:
- TOC
- CLIC
- CLOC

Instructor Activities:

1. Describe a *tactical operations center* (TOC) and its duties
2. Discuss techniques that can facilitate effective TOC operations
3. Ensure trainees understand that their TOC is "blind" and the implications this has for communicating perceived anomalies

Synopsis of Module Topics:

Introduction: A *Tactical Operations Center (TOC)* is a unit's command-and-control hub, assisting the commander in synchronizing operations. TOCs act as the primary driver of the intelligence cycle: receiving, analyzing, integrating, and distributing information across distributed teams. A TOC can *pull* information from tactical units by proactively communicating with teams and requesting specific types of information, or tactical teams may *push* relevant information to the TOC.

Small-Unit TOCs: The primary advantage of small-unit TOCs (like the Marines' CLOC) is the speed at which locally collected information can be turned into locally actionable intelligence. A battalion-level TOC has responsibility for a large area of operation, which may cover a substantial number of tactical teams. Because of this, time-sensitive intelligence may become outdated before it can be disseminated back to the tactical level. In contrast, a small-unit TOC can triage and prioritize the information it collects and rapidly identify the most valuable local intelligence.

Functionally, a small-unit TOC is similar to a traditional TOC. One difference is that the people working at the battalion level are full-time TOC staff, while the small-unit TOC contains company personnel who also have other functions. Another difference is the areas they cover. Often, it is easier to facilitate a good working relationship between a tactical team and a small-unit TOC than it is between a tactical team and a battalion operations center.

- *TOC Staff Experience:* Unfortunately, TOCs—especially small-unit TOCs—may be staffed with junior or inexperienced personnel. These relatively junior personnel often have never worked in a TOC and, as a result, are not familiar with the intricate details of TOC operations.

Charts and Visual Displays: The most effective TOCs make ample use of visual displays, such as map boards and charts. Effective visual information display techniques offer the commander a quick and easy means of getting a snap shot of his/her unit. Also, they provide the staff with an efficient means of processing information. Visual display techniques also minimize the passing of message-slips between staff sections and making numerous entries in the staff journal.

Units that have and utilize charts tend to manage large amounts of information better than those that do not. Charts alone will not make a TOC successful; its staff must first identify what critical information must be tracked. A TOC cannot process every piece of information that it receives, especially during the battle. Units must prioritize and train their personnel to distinguish between critical information and routine information. Before developing charts, consider the following:

- Determine what critical information must be tracked and displayed.
- Avoid information and chart overload.
- Build a box to store and transport charts.
- Keep a miniature version of all charts in a notebook.
- Use charts in garrison to train personnel on their use.

Recommended TOC Techniques:

- *Identify Information:* Identify and prioritize critical information to be tracked. Develop a system to track the necessary information; this system may include charts, matrices, unit symbols, or a butcher board.
- *Track Units:* Develop a system to track friendly, enemy, and civilian groups, habitual areas, and anchor points. Successful techniques include using color-coded cellophane stickers, color-coded thumb tacks, or color-coded dot-type stickers.
    - — Ensure all participants understand and use the system.
    - — Use common nicknames to describe people and places.
    - — Use precise vocabulary (such as CODIAC terminology).
    - — Develop standardized map boards so overlays can be easily and quickly transferred from map to map.
- *Minimize Chaos:* Keep the noise level in the TOC to an absolute minimum. Also, do not let the entering of messages into a journal create a backlog in the information management system. If time does not facilitate updating the journal, then keep updates in a folder and record them later.

Discussion Questions:

1. In an irregular environment, what are the benefits of speeding up the intelligence cycle?
2. In general, what challenges must TOC staff overcome? How do these challenges affect their ability to create a common tactical picture? How can distributed teams help TOC staff overcome such challenges?

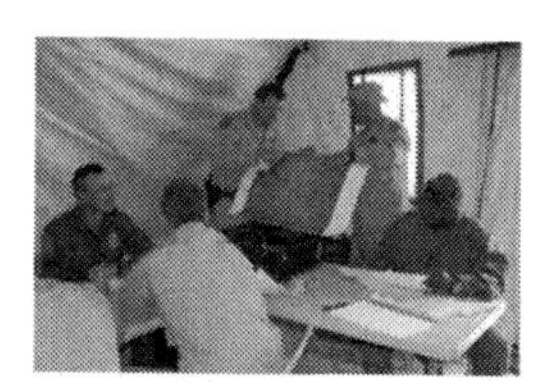

Trainees prepare their TOC during a training scenario

DID YOU KNOW?

The US Marine Corps has established *Company Level Intelligence Cells* (CLICs) and *Company Level Operations Cells* (CLOCs). These are formally established company-level TOCs.

CLICs are able to collect and analyze information and to disseminate actionable intelligence in a timely manner. The time saved by managing the intelligence cycle at the company level directly impacts the dismounted units. This allows dismounted squads and platoons to act on intelligence before it is overcome by events.

The CLOC gives a company commander the organic ability to use intelligence for increased tactical flexibility. The CLOC normally consists of watch officer and several NCOs, responsible for operations, analysis, and collection.

7004: Communicating Intelligence

Intermediate
Lecture and discussion
About 60 minutes

Terminal Objective: This module introduces trainees to the importance and methods of communicating intelligence from the tactical team up to the TOC. Upon completion of this module, trainees will be able to describe the importance of fast, secure communications within the tactical team and up the chain of command. In addition, trainees will be able to describe methods for most effectively communicating with a TOC.

Instructor Activities:

1. Describe how a tactical team should handle collected information
2. Describe how to prioritize information
3. Reiterate the difference between *information* and *analysis*
4. Describe fast, secure communication methods

Synopsis of Module Topics:

Introduction: While in the field, dismounted teams can help (or hinder) operations by effectively (or ineffectively) communicating with the TOC and lateral teams. Each team should periodically ask itself these questions:

- What do I know?
- Who needs to know?
- Have I told them?

What do I know? A tactical team can save the TOC time and effort by triaging and prioritizing the information it passes.

- *Anomalies:* When radio traffic is low, verified anomalies should be communicated to the TOC. Anomalies identified from multiple, distributed teams may create a useful intelligence picture.
- *Key Locations:* Key locations, including habitual areas and anchor points, should be communicated to and cataloged by the TOC.
- *Key People:* Key individuals within a region should be identified. In some situations (e.g., military conflicts), the whereabouts and activities of these people may be regularly communicated to the TOC. Any suspicious individuals should be identified and communicated to the TOC. Additionally, the rationale for why these individuals appear suspicious should be clearly articulated (i.e., "prove it").
- *Information Priority:* Tactical teams will collect information of different degrees of importance and time sensitivity. Teams must determine the best way to deliver this information to their TOC. Some critical information will require immediate transmission, but other data may be less time-sensitive and may be delivered by hand at the end of the operation.
- *Information vs. Analysis:* As defined previously, information is the raw data collected in the field, and analysis is the interpretation of that information. Teams should communicate information and its analyses to the TOC. However, in all cases, teams must clearly distinguish between these two components of their communications.

Who needs to know? In most cases, information should be transmitted directly to the TOC, so TOC staff can disseminate the intelligence back to other tactical teams. However, there are rare exceptions when information is so time-sensitive that one tactical team may need to communicate directly with another.

- *Suggesting Distribution:* If a distributed team knows of other teams who could benefit from their information (but who do not need to be, or cannot be, laterally contacted) the team may suggest distribution recipients to the TOC. Suggesting distribution may also be useful in situations where a tactical team cannot directly contact another unit, such as a Joint or International agency.

Have I told them? A tactical team's operations order should define a communications schedule for the team. In addition, whenever a team identifies important information or significant deviation from the baseline, it should convey that information to the TOC.

- *Secure Communications:* Tactical teams must maintain operational security during their communications. No mission, location, or unit information should ever be revealed over a radio. Specific code words for each mission should be outlined in the operations order (see the LiNDATA SitRep for example).
- *Nicknames and Vocabulary:* Efficient communication relies upon using clear, common vocabulary. Personnel should use the CODIAC terms to improve the precision of related communications. In addition, common nicknames for people, objects, and information types should be employed. Such nicknames allow tactical teams to succinctly and securely communicate with base personnel during a mission.
- *Painting a Picture for the TOC:* Remember, the TOC is "blind." TOC personnel cannot directly perceive the operational environment; hence, phrases like "he went that way" or "she ran to the left" are useless (and frustrating) to TOC staff.

Discussion Questions:

1. These communication procedures place greater responsibility on small-units and their leaders. What added duties do these recommendations place on tactical teams?
2. What are some examples of raw and analyzed information? Why is it important to communicate both types of data to the TOC?
3. In your organization, when is it appropriate to communicate with lateral teams? What CODIAC data might you communicate?

"THE OLD SAYING 'information is power' is no longer true, in the 21st Century, 'shared information is power.' How units organize their people, processes, and systems dictates the level of productivity on the battlefield of the 21st Century.
—*Exploitation Tactics (page 19)*

KEY SKILL

Working with others to construct a behavior profile of a person, event, or quarry

Instructors should emphasize the importance of developing baselines or "combat profiles" collectively, between and amongst distributed teams. Discuss specific approaches that the trainees can use to develop shared profiles.

EXAMPLE

LiNDATA SitRep: Instead of using detailed, complete sentences, LiNDATA can be communicated succinctly (and more securely), like so:

L = 12345678 (grid reference)
N = 4
D = Northeast
A = 6-8 hours
T = 1 x Panama, 1 x 6 Star Vibram

Unit 8: Conducting a Follow-up

8

If tracking is the art of reading spoor and sign, then *combat* tracking is reading spoor and sign while in a dangerous, kinetic environment. This unit focuses on application of combat tracking knowledge. Trainees will learn how to apply their knowledge and skills in realistic tactical tracking scenarios.

Suggested Prerequisites:

- CODIAC Units 1–7
- Working knowledge of how to use one's organic optics

Terminal Learning Objective:

- This unit gives trainees the opportunity to practice their combat tracking skills. These practical exercises require synthesis of all six key training objectives. At the completion of this unit, trainees will have intermediate knowledge of combat tracking, and they will possess the skills necessary to employ combat tracking in an operational kinetic environment.

Enabling Learning Objectives:

- Describe the roles and formations of a combat tracking team
- List and explain the 10 rules of combat tracking
- Demonstrate key arm-and-hand gestures
- Describe the key Lost Spoor Procedures
- Describe how to track quarry in an urban environment
- Describe key combat tracking team engagement tactics
- Demonstrate applied use of combat tracking skills in a (simulated) kinetic environment

Estimated Time Allotted for Instruction: About 38 hours

Speciality Support Requirements:

- Anti-tracking equipment for demonstration, such as a small brush and anti-tracking shoes
- Spoor Pit (preferably located near classroom) for demonstrations and a rake to clear the pit
- Ranges or other large, open locations that will support follow-up scenario-based training
- For Units 6 and 8, training participants should acquire copies of the book *Tactical Tracking Operations* by David Scott-Donelan. This publication is an important resource for learning combat tracking.

8001: The Combat Tracking Team

Introductory
Lecture and discussion
About 30 minutes

Terminal Objective: This module introduces the trainees to the duties of each member of the five person tracking team. Upon completion of this module, trainees will understand the responsibilities of each member of the team as well as the advantages of working in a cohesive tracking team.

Module Vocabulary:

- Flank Trackers
- Rear Security Tracker
- Team Leader
- Tracker

Resource DVD:

- 8001.1 – Tracking Team Overview, Video

Other Suggested Materials:

- *Tactical Tracking Operations* — Chapter 5, pages 61–62

"TRACKING BY COMMITTEE is the worst that can happen: Let the tracker do his job and everyone else do theirs."

—David Scott-Donelan

Instructor Activities:

1. Define and discuss the roles and duties of the tracking team
2. Describe and discuss the advantages of the combat tracking team
3. Review the Tracking Team video

Synopsis of Module Topics:

Introduction: Combat tracking is a team effort. The five-person team provides mutual security and extra eyes, and minimizes fatigue.

Team Leader: The *team leader* is responsible for the tracking team. He/she makes the decisions regarding the tracking operation, tactical decisions, navigation, communications, security, and the conduct of the team. The team leader is the only position that does not rotate through the course of the tracking operation.

Tracker: The *tracker* is responsible for following the spoor and keeping the team leader aware of any intelligence he/she determines from the spoor (e.g., number and speed of the quarry or any changes that occur with the quarry).

Flank Trackers: There are two *flank trackers* who are primarily responsible for protecting the team from ambush. Flank trackers should continually scan the terrain ahead, looking for any indication of a hostile force. In addition, they can be instrumental in lost spoor procedures, reconnaissance, and tactical recommendations.

Rear Security Tracker: The *rear security tracker* is responsible for a variety of tasks as needed by the team leader. Based upon the tactical situation he/she can operate equipment (such as GPS or radio), provide rear protection or forward reconnaissance, back up the tracker, assist with lost spoor procedures, or perform any other duty as required by the team leader.

Advantages: The five-person team is self-contained and provides its own protection. The team can also track faster, collect information quicker, and re-locate lost spoor more effectively that a single person alone. Each team member should be cross-trained, so positions can be rotated to prevent fatigue.

Discussion Questions:

1. Does the five-man tracking team fit well with your existing team or unit sizes? If not, how can you adjust the five-man model to better work for your organization?
2. How might you use a six people? What role would you double-up?

8002: Combat Tracking Team Rules

Instructor Activities:

1. Discuss the ten primary rules of tracking
2. Review the Rule video
3. Discuss the potential consequences for failing to follow the rules

Introductory
Lecture and discussion
About 60 minutes

Terminal Objective: This module introduces trainees to the 10 rules of tracking. Upon completion of the module, trainees will be able to recall and describe the importance of the rules of tracking.

Resource DVD:

- 8002.1 – Rules, Video

Other Suggested Materials:

- *Tactical Tracking Operations* — Chapter 3, pages 33–34

Synopsis of Module Topics:

Introduction: These rules have been developed over time by trackers through real world experience. Many of the rules appear obvious, but there is always someone who has not followed them. And, like most rules, not following these rules can lead to failure.

Rules of Tracking:

1. *Correctly identify the tracks you wish to follow.* Valuable time can be lost if the tracking team does not identify the correct tracks.
2. *Mark and record map coordinates of the start point which is called the Initial Commencement Point (ICP).* This is important both for the combat tracking team and for outside communications.
3. *Never walk on top of ground spoor.* The tracking team should try not to disturb ground spoor; it may be used again to backtrack, etc.
4. *Never overshoot the Last Known Spoor (LKS).* This is the most important rule in tracking. Failure to do so will waste valuable time.
5. *When following aerial spoor, always check for confirmable evidence.* Aerial spoor is less reliable than ground spoor, and a tracker should always look to confirm it with some other spoor or sign.
6. *Always know exactly where you are.* It is vital that the combat tracking team always know its location.
7. *Always keep in visual contact with other team members.* If visual contact is not maintained, members cannot use hand signals.
8. *Always try to anticipate what your quarry will do.* An experienced tracker can get into the mind of the quarry and anticipate their moves.
9. *The tracker sets the pace of the follow-up.* The tracker will inevitably be the slowest; the other members must conform to his/her pace.
10. *Never "force" a track to conform with your own preconceptions.* It is easy for a tracker to "see" what he/she expects, instead of reality.

Discussion Questions:

1. If you could add one more rule to the list, what would you include?
2. Is it ever appropriate to break a rule? Why or why not?

8003: Tracking Team Formations

Introductory
Lecture and discussion
About 60 minutes

Terminal Objective: This module introduces trainees to the tactical techniques and formations used by a combat tracking team. Upon completion of this module, trainees will be familiar with the four visual tracking formations.

Module Vocabulary:

- Extended Line Formation
- Half Y Formation
- Single/Ranger File Formation
- Y Formation

Resource DVD:

- 8003.1 – Tracking Team Formations, Video
- 8003.2 – Tracking Formations in Practice, Video

Other Suggested Materials:

- *Tactical Tracking Operations*
  — Chapter 5, pages 62–65
  — Chapter 5, pages 69–71

Instructor Activities:

1. Describe the tracking team formations and factors affecting their use
2. Discuss the importance of maintaining tracking team security
3. Discuss the need to maintain noise discipline and visual camouflage
4. Discuss the benefits and limitations of running on spoor
5. Review the two videos provided on the Resource DVD

Synopsis of Module Topics:

Maintaining Team Security: The combat tracking team is a small unit that must often operate in a hostile or non-secure environment. To minimize the chances of being seen or heard prematurely, the tracking team must continually maintain high standards of *tactical movement, camouflage*, and *noise discipline*. Considerable training, discipline, and control are necessary on the part of the entire team to ensure that they maintain the tactical advantage over their adversary.

Tactical Movement: Appropriate tactical formations are important. In most cases, the terrain will dictate the type of formation a tracking team adopts. Other factors affecting the choice of team formation include:

- Visibility through the vegetation traversed
- Terrain and ground conditions
- The time/distance gap
- The prevailing or oncoming weather conditions
- The tactical situation

Team Formations: The Y formation represents the standard tactical tracking team formation. It can be modified for specific terrain or tactics.

- *Y Formation:* The basic Y Formation can be used in open or lightly wooded areas. The distance between team members varies depending upon the visibility, which is affected by the terrain, vegetation, and lighting. The tracker is positioned on the spoor; the flank trackers are at tracker's front approximately 45° off the line of the track, one to each side; the team leader follows behind the tracker, and finally, the rear security tracker follows the team leader. This formation provides tactical and tracking flexibility. It allows the flank trackers to observe the terrain to their front and sides, looking for any signs of hostile forces, ambush, or spoor. If the track veers off to the side, one of the flank trackers will often identify the new track before the tracker. When this occurs, the team can rotate around, so the flank tracker becomes the tracker and the original tracker takes his/her place on the flank.

- *Half Y Formation:* The half Y formation is used in place of the Y formation when the quarry's trail runs alongside an obstacle, such as a river bank, rock face, or lake. The flank tracker on the side of the obstacle moves back behind the tracker or team leader and continues to scan to that flank from his/her new location.
- *Single/Ranger File:* The single file formation (sometimes called ranger file) is used in thick vegetation and other areas of limited visibility. The flank trackers position themselves behind the tracker or bracketing the team leader. It is imperative that even in limited visibility the team remains in visible contact at all times.
- *Extended Line:* The extended line formation is used in wide open areas with little or no danger of ambush. The tracker is again positioned on the spoor. The team leader and rear security stand on each side about 20 feet away. The flank trackers are positioned on the far ends of the line another 20 feet away. This formation covers the most amount of ground and is also useful if the tracks are difficult to see since each team member can assist the tracker.

Camouflage: Camouflage plays an integral role in the combat tracking team, both to conceal the team and to increase the team's awareness of the terrain around them. Team members should be appropriately camouflaged to the terrain type.

Noise Discipline: As a small unit operating in close proximity to hostile forces, it is important for a combat tracking team to practice noise discipline. While tracking quarry, hand signals should be the primary form of communication between team members.

Running on Spoor: Tracking teams may be inclined to run on the track line in order to close the time/distance gap. However, there are drawbacks that must be carefully considered before the team runs on the spoor.

Discussion Questions:

1. Why do you think the five-person tracking formation is recommended? How would you organize a four-person or six-person tracking formation?
2. What are some pieces of gear that might be visible to your quarry? How can you help minimize the visibility of your own binoculars, glasses, sunglasses, and other gear?
3. How far do you think the scent of cigarette smoke travels? How about the scent of cologne or body sweat? The sound of talking? The hiss and beep of a radio?

To do their jobs effectively, tracking teams have to break the cardinal rule of small unit operations, which is "*Never go where you can be expected.*" Tracking teams must put themselves right into the face of the enemy, but must do so with the utmost level of precaution to avoid unnecessary casualties. Movement, personal and team camouflage, silence, and the correct team formation commensurate with the ground and terrain are essential components of team security and must be monitored constantly while the team is on follow-up.

—*David Scott-Donelan*

REMEMBER!

Your team control is inversely proportional to your speed. Be particularly cautious when running on the track line!

8004: Arm-and-Hand Signals

Introductory
Lecture and discussion
About 60 minutes

Terminal Objective: This module introduces trainees to the hand-and-arm signals used within the tracking team for silent communications. Upon completion of this module, trainees will be able to describe and demonstrate the key hand and arm signals.

Resource DVD:

- 8004.1 – *Army FM 21-60*
- 8004.2 – Signals, Video
- 8004.3 – Trainees in the Field, Video

Other Suggested Materials:

- *Tactical Tracking Operations* — Chapter 5, pages 66–67

Instructor Activities:

1. Discuss the advantages of clear, silent communications
2. Review the two videos provided on the Resource DVD
3. Demonstrate the key hand-and-arm signals for the tracking team

Synopsis of Module Topics:

Introduction: Silent hand-and-arm signals are the means whereby members of a tracking team communicate with each other, without giving forewarning to a hidden or stationary quarry. While conducting a follow-up the team members must remain in visual contact with each other to ensure hand-and-arm signals can be observed. It is especially important for the flank security to regularly look at the tracker and team leader to maintain visual contact.

Demonstrate the following signals that can be used while tracking:

- *All Clear:* Indicated by the "thumbs up" signal; the hand is held a shoulder height in front of the body, with fingers tightened into the palm and the thumb extended upward
- *Change of Direction:* Signaled by extending the arm outward to the side of the new direction, palm forward
- *Come in:* Signaled by raising the arm vertically, palm to the front, and waving in large horizontal circles.
- *Come Talk to Me:* Signaled by pointing toward the team member(s); beckon by holding the arm horizontally to the front, palm up, and motioning toward the body
- *Commence a 360:* Signaled by using the index finger to draw a large, vertical circle in front of the body
- *Commence a Cross-Over:* Signaled by holding the arms above the head, with wrists crossed directly above the head and open palms facing forward; then both arms are moved out, away from the body, until the hands are just above shoulder height
- *Extend into Line:* Signaled by holding both arms straight out to the sides, at shoulder height, forming one continuous horizontal line
- *Go Look:* Point in the direction that should be observed, followed by pointing the index finger to one eye and middle finger to the other.
- *I Am Scanning with My Weapon Mounted Scope:* Signaled by holding one's weapon normally, looking through the scope, but also holding up the forward supporting hand with the palm facing to the rear and supporting the weight of the weapon with the bend of the elbow

- *Lost Spoor Search:* Signaled by by the team leader to flank trackers when he/she wants them to initiate a search; the team leader points his/her index finger upward, revolving the hand, and pointing in the direction he/she wants searched
- *Move Forward:* Signaled by facing in the direction of movement, raising the hands to should level, palms to the front, and then moving the hands forward and backward
- *Move Out:* Signaled by the team leader; he/she faces in the direction of movement, holds his/her arm extended to the rear and swings them overhead and forward in the direction of movement, palm down
- *Obstacle Ahead:* Signaled by holding an open hand out in front of the body, with the palm facing toward the face, and then bringing it directly in front of the face
- *Open Radios:* Signaled by raising the hand to the ear with the thumb and little finger extended
- *Quarry Seen:* Signaled by any team member who spots the quarry; the team member freezes, then slowly raises his/her weapon to the shoulder, pointing it at the quarry; all team members repeat the signal when they have the quarry in sight
- *Spoor Found:* Indicates a team member has relocated the spoor; signaled by making the opposite of the "spoor lost" signal: the hand, with the fingers extended, is pointed down toward the relocated spoor
- *Spoor Lost:* Signaled by the tracker when he/she temporarily loses the spoor; signaled by holding out the hand, palm forward with the fingers extended and pointing directly upward
- *Take It:* Signaled by holding the arm straight out to the side at shoulder height, then the hand is lowered with open fingers and raised to mimic the action of picking up an object and "handing it" away.
- *Track Trap Ahead:* Signaled by flank tracker to the team leader by opening and closing his/her hand, like a puppet's mouth.

Lost Spoor Signal

Eyes on Target Signal

All Stop Signal

Discussion Questions:

1. Why are silent signals superior to radio communications during a follow-up? When should radios be used during a follow-up?
2. How will the team know if the rear security is signaling? Why might the rear security need to signal to the team?
3. What other covert ways might a tactical team use to communicate with each other during a follow-up? Why might these be useful?

8005: Lost Spoor Procedures

Intermediate
Lecture and discussion
About 120 minutes

Terminal Objective: This module introduces the formal lost spoor procedures used during a follow-up. Upon completion of this module, trainees will be able to discuss the employment of lost spoor procedures.

Module Vocabulary:

- Alleyway Scans
- Bitterman Technique
- Box Search
- Cross-Over
- Flanker 360
- Last Known Spoor (LKS)
- Lost Spoor Procedures (LSP)
- Quick Scan
- Track Trap
- Tracker 360

Resource DVD:

- 8005.1 – Lost Spoor, Video

Other Suggested Materials:

- *Tactical Tracking Operations* — Chapter 3, pages 46–50

Instructor Activities:

1. Discuss how spoor can be lost and importance of relocating it
2. Review the Lost Spoor video
3. Discuss and demonstrate the *quick scan* technique
4. Discuss and demonstrate formal *lost spoor procedures*
5. Discuss and demonstrate short cuts for lost spoor procedures
6. Discuss the importance of team discipline during these procedures

Synopsis of Module Topics:

Lost Spoor Procedures: It is very common for trackers to lose the spoor at some point during a track. The cardinal rule for tracking is never advance beyond the Last Known Spoor (LKS). When executed properly, lost spoor procedures will relocate the spoor in 90% of the cases.

Quick Scan: An informal quick scan procedure can be initiated before signaling the team that the spoor has been lost. When one can no longer see the spoor ahead, stop and carefully scan the terrain in a 240° arc.

Formal LSPs: When a tracker can no longer see spoor in the terrain ahead, he/she should alert the team with the hand signal, and then mark the LKS. At this point, the tracking team can initiate Lost Spoor Procedures (LSPs).

- *Likely Lines:* After the LKS has been marked, the tracker searches ahead for what the most likely path the quarry has taken. The tracker repeats this procedure until he/she has exhausted all of the likely lines.
- *Tracker 360:* In the tracker's 360 procedure, the tracker moves to a spot about 20 yards behind the LKS and then walks in a circle around the LKS but within the protection of the flank trackers. The tracker searches the ground and vegetation carefully while positioning the light to maximum advantage. If the spoor is not found, the team leader can ask the tracker to conduct another 360, this time outside the physical protection of the flank trackers but within the protection of effective weapons range.
- *Flanker 360:* Depending on the security situation, the flank trackers can conduct interlocking 360 degree searches out ahead of the tracking team to try and relocate lost spoor.
- *Box Search:* In a box search, the team tries to determine natural or artificial lines such as roads, fences, rivers, or railroad tracks that surround the area where the spoor was lost. Once these lines are determined, the team searches around the perimeter of where the tracks were lost.

Short Cut LSPs: In addition to the formal LSPs, these short cuts may be implemented at any time. Short cuts can also be implemented while the tracker is still following the spoor.

- *Cross-Over:* If the quarry has been maintaining a fairly straight track line, then the team leader can implement the cross-over technique. In this procedure, the two flankers "cross over" to the other side of the formation while simultaneously searching the ground for the spoor.
- *Track Trap Search:* The flankers often see prime tracking terrain features ahead of the team, such as a dirt road, a sandy patch, cultivated field, or earth bank. These "track traps" may hold indicators, if quarry passed that way. Upon seeing a potential track trap, a flanker should signal to the team leader, and then if approved, one or both of the flankers should move along the "trap" while checking the ground for spoor.
- *Bitterman Technique:* The Bitterman Technique is used when the quarry is marching on an azimuth. The team leader lines the team along the track line and takes the azimuth with a compass, and then after roughly one-half mile, he/she repeats the process. If the two azimuths are identical then the conclusion can be made that the quarry is maintaining direction by the use of a compass or GPS. In this instance the team leader should extend the azimuth on his/her map, attempting to identify a landmark toward which the tracks are heading.
- *Alleyway Scans:* Alleyway scans are used when the spoor moves along a used track, trail, riding path, or stream bed. The team leader sends one of the flank trackers out about 50 yards along the path and the other 25 yards along on the other side of the path. Each flank tracker then examines the trail for spoor. If no spoor is found the Flanks move another twenty-five yards ahead along the trail and repeat the process.

Team Discipline: Although it may seem that employing all team members in LSPs would be efficient, this often slows the team down due to contamination. The team must exercise tactical patience while the tracker, and sometimes the flankers, search for lost spoor.

"TWO THINGS ARE A 'GIVEN' in tracking armed and dangerous humans, whether they are fugitives from justice, insurgents or enemy soldiers. The first is the absolute necessity to close the time and distance gap and the second that the spoor, despite all efforts to the contrary, will be lost at times."

—*David Scott-Donelan*

Discussion Questions:

1. Is it a good idea, or a bad idea, for the other team members to start looking for the spoor when the tracker signals that it is lost? Why or why not?
2. What examples of track traps can you think of?

8006: Macro-Tracking (Activity)

Intermediate
Practical application
About 6 hours

Terminal Objective: This module gives trainees the opportunity to practice their team formations and arm-and-hand signals. Upon completion of this module, trainees will be able to demonstrate effective use of macro-tracking team formations, communications, and tracking team coordination techniques under basic conditions.

Module Vocabulary:
- Macro-tracking

Resource DVD:
- 8006.1 – Field LSP, Video

REMEMBER!
A tracker should look as far ahead as the spoor is visible and move toward that point while scanning ahead for more indicators. The tracker should avoid looking too close to his/her own feet because it will slow down the tracking team, impede the tracker's awareness, limit his/her visual contact with the team, make the tracker more prone to ambush, and result in a higher degree of eye strain.

Instructor Activities:

1. Preplan the day's activities with the staff
2. Brief the class on the day's activities
3. Review the Field LSP video
4. Divide trainees into their 5–6 member tactical teams
5. Allow teams to "leap-frog" follow-ups during the practice exercises
6. Conduct team after-action reviews after each small exercise

Synopsis of Module Topics:

Preparation: Trainees will divide into their small teams and conduct several follow-ups on a relatively easy terrain. No role-players or other special equipment will be needed by the instructors; however, trainees will need communications equipment and their full kits, including optical devices. Instructors should also have access to their own "instructor net" radios (or cell phones if radios are not available).

Transportation: Trainees and instructional staff should be bussed to their respective follow-up locations. Two teams and at least one member of the instructional staff should be dropped off at each designated start location. The start locations should be sufficiently spread-out so that teams do not run into each other, and a central "watering hole," pick-up locations, and exercise end time should be carefully pre-planned.

Team Briefings: The pairs of teams will take turns laying spoor and then conducting follow-ups. Instruct follow-up teams to take about a 10 minute head-start, and then stop when a hide is found around 15–30 minutes later. As the first team lays spoor, the instructor should remain with the follow-up team and discuss relevant tracking techniques and tactics. The instructor should remind the team to practice good team formation tactics, use their arm-and-hand signals, maintain security, use their optics to good effect, and use appropriate Lost Spoor Procedures.

Follow-Up: At least one member of the instructional staff should accompany the follow-up team to observe their performance. Use the tracking Behavioral Observation Checklist, found on the Resource DVD, to assist with this performance assessment. Special attention should be paid to the team's communication, coordination, formations, security protocols, and use of Lost Spoor Procedures.

After-Action Review: After completing the follow-up, the two teams should briefly debrief and then switch roles so that the follow-up team becomes the spoor layers. Repeat this leap-frogging until all trainees have filled the roles of a *tracker* and a *team leader* at least once.

8007: Macro-Tracking 2 (Activity)

Instructor Activities:

1. Preplan the day's activities with the staff
2. Brief the class on the day's activities
3. Divide trainees into their 5–6 member tactical teams
4. Role-players (or instructional staff) lay spoor trails
5. Brief each team on their scenarios
6. Allow each team to conduct their follow-ups
7. Conduct an after-action review with each team

Synopsis of Module Topics:

Preparation: Trainees will divide into their small teams and conduct two or three follow-ups on an intermediate-level terrain. Role-players should lay the spoor for each follow-up. One or more role-players should be assigned to each trainee team. During the preparation phase, the instructors should help the role-players learn relevant scenarios, such as acting as a group of drug traffickers or a kidnapper with hostages. During this planning, consider what sort of equipment (to make ground spoor impressions with) and/or loose litter (to drop) the role-players can carry.

Transportation: Role-players should be taken to the tracking exercise area first so that they can begin laying spoor about 30 minutes before the trainees arrive. Trainees and instructional staff should be bussed to their respective follow-up locations, next. The 5–6 person teams should be sufficiently spread-out to avoid running into each other, and a central "watering hole" and pick-up locations should be carefully pre-planned.

Team Briefings: Each team should be briefed on their first scenario. Instructors should encourage trainees to exercise tactical patience and discipline, as well as careful security procedures.

Follow-Up: At least one member of the instructional staff should accompany each tactical team to observe their performance. Use the tracking Behavioral Observation Checklist, found on the Resource DVD, to assist with this performance assessment. Special attention should be paid to the team's communication, coordination, formations, security protocols, and use of Lost Spoor Procedures. Role-players should be encouraged to react realistically to the trainees (e.g., armed fugitives should try to ambush the trainees if the tracking team is seen coming); however, role-players should not work too hard to circumvent the trainees (yet).

After-Action Review: As before, after each follow-up the instructor should debrief the team (while the role-players lay new spoor).

Intermediate
Practical application
About 6 hours

Terminal Objective: This module is similar to the previous one, except that this module uses scenario-based training, which will help trainees begin to integrate their knowledge and skills, making these abilities more automatic and reinforcing effective decision-making strategies. Upon completion of this module, trainees will be able to demonstrate effective use of macro-tracking team formations, communications, and tracking team coordination techniques—and to do so under realistic operational conditions.

"YOU WILL NOT CATCH your quarry if you aren't constantly thinking about closing the time-distance gap"

—*David Scott-Donelan*

8008: Missions Supported by Tracking

Introductory
Lecture and discussion
About 60 minutes

Terminal Objective: This module introduces the operational procedures associated with conducting a follow-up. Upon completion of this module, trainees will be able to clearly describe the activities that occur during a typical follow-up, and they will be able to describe alternative follow-up missions, including back tracking and urban tracking procedures.

Module Vocabulary:

- Back Tracking
- Night Tracking
- Spoor Cards
- Urban Tracking

Other Suggested Materials:

- *Tactical Tracking Operations*
  — Chapter 3, pages 50–52
  — Chapter 5, pages 71–74
  — Chapter 6

"Tracking is one of the best sources of immediate use intelligence, information about the enemy that can be put to use immediately."

—The US Army FM 17-98, Scout Platoon manual

Instructor Activities:

1. Discuss the intelligence tracking teams *must* gather
2. Discuss the intelligence tracking teams *should* gather, if available
3. Discuss alternative follow-up mission activities

Synopsis of Module Topics:

Introduction: Combat tracking skills augment a variety of operations:

- Pursuit to contact
- Locate arms caches
- Recovery of wounded personnel
- Counter surveillance
- Information and intelligence collection
- Combat Search and Rescue (Air Force CSAR)
- Maintain contact with a fleeing enemy after a firefight
- Back-tracking to source
- Enemy route and infiltration investigation
- Sensor sites and placement
- Counter drug operations
- Area interpretation and analysis
- Forensic analysis
- Border Patrols (assess corridors and routes)
- Clandestine operations (movement to recon and sniping hide sites)
- Locating enemy mortar and rocket firing sites

Typical Follow-Ups: Unless a combat tracking team discovers tracks while on patrol, follow-up mission orders typically come from their immediate senior command element. Typically, once this command element instigates a tracking mission, the tracking team travels to the Initial Commencement Point (ICP) and reports a LiNDATA SitRep. Each set of relevant prints should then be nicknamed and visually recorded, either on a *spoor card* or via a photo. If not already interviewed, witnesses at the ICP should be asked about the quarry—although these reports must be considered carefully, since witnesses often intentionally or unintentionally report false information. After these activities are complete, the team can begin following the track line with the goals of (1) closing the time/distance gap using ***all*** aggressive means available and (2) attempting to get into the mind of the enemy.

Night Tracking: Nighttime follow-ups can be conducted using Infrared/IR lamps, LEDs, incandescent lights, or light sticks.

Back Tracking: Back tracking involves following a trail back to its source in order to gain useful intelligence. Through back tracking, a team can determine where the quarry originated and may be able to determine other sites of importance to the quarry. In a law enforcement environment, this could mean finding other drug storage or growing areas. In a military environment, back tracking could lead the tracker to operational bases or hideouts. The biggest drawback with backtracking is the drain on manpower. That is, a second tracking team is required to conduct the backtrack while the main team conducts the follow-up.

Tracking with canine assistance

Reconnaissance and Surveillance: Reconnaissance involves general overwatch of a region, while surveillance is specific observation of a person, place, or a suspect. Tracking can be employed in such missions to follow a quarry that leaves the observed area or to collect additional intelligence about the enemy's movements within an area of operations.

Urban tracking

Urban Tracking: While more difficult, it is still possible to track quarry through an urban environment. Since people are generally lazy, the quarry will take short cuts. This lets trackers determine likely lines and use lost spoor procedures to follow the spoor. Since much of urban tracking requires the repeated use of lost spoor procedures, the five-man team is more engaged in actual tracking than security. Therefore it is important in urban environments to provide the combat tracking team with an additional security detail. If sufficient security is available, the team can divide individual trackers, each with his/her own security. Beginning at the Initial Commencement Point (ICP) the tracking team breaks into pairs with each pair moving along the road and intersecting streets and alleyways searching for track traps along likely avenues of outlet for the quarry. When a cross street is reached, each pair covers and searches each of the three possible directions (straight, left, and right) for an indicator. When an indicator is found, the team moves forward to the next cross street and repeats the process.

Tracking and surveillance

Discussion Questions:

1. How could tracking skills be used to support clandestine operations? How about sniper teams?
2. How might urban tracking support contemporary military or law enforcement missions?
3. Think about your own occupation; with what kinds of operations could tracking skills assist? What field operations would *not* be aided through the use of tracking skills?

8009: Urban Tracking

Intermediate
Lecture and discussion
About 60 minutes

Terminal Objective: This module provides trainees with the opportunity to practice their urban tracking skills. Upon completion of this module, trainees will have applied knowledge and the ability to effectively conduct basic urban tracking operations.

KEY SKILL

Making effective/efficient identification of anchor points and indications of anti-tracking

Instructors should remind trainees to think about how their quarry moves and watch for cues in the terrain that indicate the enemy's mindset. Is the quarry moving toward an anchor point? Is the quarry using anti-tracking?

Instructor Activities:

1. Remind trainees how to anticipate the path of their quarry
2. Remind trainees to use all assets available to them
3. Discuss tactics for urban tracking operations

Synopsis of Module Topics:

Introduction: Urban tracking was introduced in the previous module. During urban tracking operations, personnel will need to make constant use of *alleyway scans*, looking for where their quarry's trail diverges from a hard surface or contaminated area.

Use all Assets: When conducting urban tracking operations, a dismounted team must use all assets available to them. This will give them an advantage over their quarry, despite being on less-than-advantageous tracking terrain. These advantages include:

- *Get into the Mind of the Quarry:* Remember, most people are creatures of habit; they are generally lazy and take short cuts whenever possible. Look for likely lines, short cuts, habitual areas, and anchor points. Think about landmarks and other locations of interest that the quarry might head towards or at which they might stop.
- *Look for Track Traps:* Look for track traps near likely landmarks or the edges of roads/paths to verify the quarry's track line. Track traps can be found in unexpected places, including the edges of roads, gardens or freshly cut grass, dewy grass, steep inclines or embankments, dust along the side of roads, oil patches or dirt in parking areas, puddles or mud patches, or areas of fallen leaves.
- *Look for Sign:* Never forget to look for sign. What did the quarry drop? What did the quarry disturb as he/she passed? Do you hear dogs barking? Are the atmospherics of an area disturbed?
- *Ask Witnesses:* Often in urban areas many witnesses are available to interview. Unfortunately, witnesses often provide inaccurate or intentionally false information. However, they can sometimes provide valuable intelligence. As appropriate ask passersby if they have seen your quarry or anything outside of the ordinary (related to your quarry).
- *Use Your Manpower:* Generally, military and law enforcement agencies have greater manpower than the quarry that they are pursing. Make use of this advantage by leap-frogging teams, placing one or more small units at likely landmarks in the direction of the quarry's travel. These teams can scout for tracks and prepare to intercept the quarry if he/she travel in that direction.

8010: Macro-Tracking 3 (Activity)

Intermediate
Practical application
About 6 hours

Terminal Objective: This module provides trainees with the opportunity to practice their urban tracking skills. Upon completion of this module, trainees will have applied knowledge of and the ability to effectively conduct basic urban tracking operations.

Instructor Activities:

1. Prepare any handouts (e.g., photos) that will be given to trainees
2. Role-players (or instructional staff) lay spoor trails
3. Divide trainees into their 5–6 member tactical teams
4. Brief each team on their scenarios
5. Allow each team to conduct their follow-ups
6. Conduct an after-action review with each team

Synopsis of Module Topics:

Preparation: Instructors must preplan which routes each role-player or instructional staff member will take. The trainees will later follow these paths, so they must be carefully selected. During the preparation phase, the instructors should also determine the scenarios behind each of the follow-ups. Scenarios should relevant to the trainees' operational environment. Example scenarios might include following an armed fugitive who just left an robbery, following a suspicious man seen running from an IED explosion who may or may not have been involved, or even tracking down a lost and disoriented elderly person. As appropriate, instructors should create photographs, sketches, or pre-written "witness statements" to supplement the scenario briefings.

Team Briefings: After trainees are divided into the 5–6 member tactical teams, each team should be briefed on their scenario, the Initial Commencement Point (ICP), and general direction in which the quarry was seen leaving the scene. Encourage trainees to use all resources available, including asking passersby whether they have seen the quarry and making use of their personal knowledge of the area to identify likely landmarks that the quarry might be traveling towards.

Follow-Up: At least one member of the instructional staff should accompany each tactical team to observe their performance. Use the tracking Behavioral Observation Checklist, found on the Resource DVD, to assist with this performance assessment. Special attention should be paid to the team's communication, coordination, formations, and security protocols, as well as to whether the trainees considered possible anchor points, habitual areas, or short cuts. Instructors should also consider what scenario-specific tactics are most appropriate and whether trainees considered these tactics.

After-Action Review: After completing the follow-up, each team should engage in an individual debrief with a member of the instructional staff, and then the whole class should be reconvened to participate in an overall discussion and debrief.

For this exercise, trainees can be given a photo of their quarry, as well as a photo of their quarry's prints. These simulate the type of eye-witness information that may be available during urban tracking, and it will facilitate interactions with other people in the urban environment.

8011: Anti/Counter Tracking

Introductory
Lecture and discussion
About 60 minutes

Terminal Objective: This module introduces the trainee to the various types of anti-tracking techniques that the quarry can employ in order to escape and evade trackers. Upon completion of this module, trainees will be able to identify and overcome anti-tracking techniques used against them. In addition, trainees will be able to describe and discuss the advantages/disadvantages of each technique.

Module Vocabulary:

- Anti-Tracking
- Counter Tracking
- Spoor Reduction Techniques

Resource DVD:

- 8011.1 – Anti-Track, Video
- 8011.2 – Anti-Tracking Spoor Pit Demonstration, Video

Other Suggested Materials:

- *Tactical Tracking Operations* — Chapter 7

Instructor Activities:

1. Discuss the techniques a quarry can employ to escape the tracker
2. Discuss what the tracker can learn from anti-tracking techniques
3. Review the two videos provided on the Resource DVD

Synopsis of Module Topics:

Introduction: A quarry may try to employ anti-tracking techniques to escape and evade any pursuit. There are four methods used to fool, counter, or harm a tracker or tracking team:

- *Increase speed and distance* is an effort by the quarry to outpace the tracker so that the tracker cannot close the time/distance gap. From the quarry's perspective it has the advantages of:

  — Effective against unskilled or search and rescue trackers
  — Effective when escape distance is short
  — Less time for the quarry in the area of operations
  — Escape may take place before their presence is detected

  For the tracker, it has the advantages of:

  — Speed creates more easily recognizable spoor
  — Speed increases the possibility of the quarry having a problem
  — Speed increases noise and reduces awareness
  — Can be countered by leap-frogging teams once direction is known
  — Fatigue and the need for water increase
  — Difficult to employ with large or heavy loads
  — Difficult to maintain in hot arid conditions
  — Overall speed is reduced to that of the slowest person
  — Wounded quarry become a hindrance
  — Increases mental pressure on quarry
  — Tracks linger longer, increasing their likelihood of being found

- *Anti-tracking* techniques are used by the quarry to disguise or conceal the spoor in an attempt to fool the tracker. From the quarry's perspective these techniques have these advantages:

  — Will slow down less experienced trackers
  — Gives the quarry a temporary increase in the time/distance gap

  For the tracker, they have these advantages:

  — Slows down the quarry each time it is used
  — Means extra work for the quarry
  — Alerts and educates the tracker each time it is used
  — Some techniques may play into the hands of trained trackers

- *Spoor reduction techniques* allow a quarry to scatter or split up making the spoor progressively more difficult to follow. Techniques employed depend upon the size of the group, the distance to be covered and presence of a hostile local population. The three techniques are:

  — *Break-away groups*: Splitting up into smaller groups.
  — *Bomb-shelling*: Each member of the quarry leaves the contact area in a different direction, usually to rally at a prearranged point later.
  — *Drop-offs:* This technique involves the group leaving together, but some time later individuals or pairs leave the main group. This is one of the more effective techniques, especially when the drop-offs occur in areas where the ground is hard and spoor is difficult to pick up.

  From the quarry's perspective, the advantages are:

  — Decreases motivation of the tracking team
  — Easy to employ and does not slow insurgents
  — Overworks tracking teams
  — Difficult to identify correct spoor
  — Increases time/distance gap

  For the tracker, the advantages are:

  — Quarry must have clearly defined rally points
  — By splitting up, individuals become vulnerable
  — Time taken at the rally point increase time in the area of operations
  — Time at rally point gives time for discovery
  — Time to assemble depends on slowest person
  — Increase chances of the quarry being spotted
  — Group firepower is reduced
  — Unless the quarry has radios, communication is reduced

- *Counter tracking methods* are employed to harm pursuing trackers or do psychological damage, making them lose interest in the follow-up. Typical examples include ambush, booby traps, or fire. From the quarry's perspective the advantages are:

  — Casualties cause psychological damage
  — Casualties require time for medical evacuation
  — Can deter less trained or motivated trackers
  — Detonating traps indicate trackers' location

  For the tracker, the advantages are:

  — Setting traps is dangerous and time consuming
  — Casualties motivate well-trained teams
  — Traps confirms quarry's intent to evade capture

DID YOU KNOW?

A variety of anti-tracking gear can be made or purchased.

Footwear: Anti-tracking footwear is fairly common. For example, along the US–Mexico line, individuals attempting to cross the border illegally may tie carpet mats and foam to their feet, to disguise their prints. Of course, more sophisticated anti-tracking boots, or boot covers, can be purchased.

Brush: A simple feather-duster or brush can effectively obliterate tracks; however, this technique is time consuming. A more rapid version of this approach has been used by illegal border crossers: They use leaf-blowers to eliminate tracks. Although there are obvious limitations (noise, bulk, etc.), even a leaf-blower can be an effective anti-tracking tool.

David Scott-Donelan demonstrates anti-tracking techniques in a spoor pit

8012: Tracking Team Engagement Tactics

Intermediate
Lecture and discussion
About 60 minutes

Terminal Objective: This module introduces trainees to the engagement tactics employed by a dismounted tracking team if/when they engage their quarry. Upon completion of this module, trainees will be able to recall and describe the three basic types of engagements and their team's appropriate tactical response to each.

Module Vocabulary:

- Cover Shoot
- Encounter Action Drill

Other Suggested Materials:

- *Tactical Tracking Operations* — Chapter 5, pages 68–69

Instructor Activities:

1. Discuss the three types of engagements that can occur
2. Discuss the team's tactical response to each engagement
3. Remind trainees of the concept of situational awareness
4. Discuss the *cover shoot* technique
5. Discuss the importance of being familiar with one's role

Synopsis of Module Topics:

Introduction: The goal of combat tracking is to close the time/distance gap with the quarry. Therefore, in many instances, the team who conducts the tracking will not engage the quarry but will instead call in a separate strike team. However, dismounted teams must always be prepared to engage—or be engaged by—the enemy.

Engage the Quarry: Generally, there are three situations that will lead to a tactical engagement between a tracking team and their quarry:

- *Team sees quarry first:* When a combat tracking team member sees the quarry, the Encounter Action Drill is to freeze and slowly bring one's weapon up to one's shoulder, pointed at the quarry. This action alerts the team to the position of the quarry and also puts at least one team member in position to immediately fire upon the quarry. Following this, the team should execute the appropriate tactical engagement based upon its rules of engagement. *Do not* dive into cover when the quarry is spotted; this movement will alert the enemy.
- *Team and quarry see each other simultaneously:* This tactical situation will occur frequently; when it occurs, the tracking team must take immediate control of the situation by advancing aggressively on the quarry with weapons raised and in the firing position. The team leader should challenge the quarry to drop their weapons and raise their arms. If the quarry complies, the team can continue to advance and conduct normal disarming and capture procedures. If the quarry does not comply, then the team is in position to initiate an immediate ambush.
- *Quarry ambushes the team:* Any dismounted team could be ambushed, and team members should always maintain awareness of potential ambush sites. A standard tactical response for a unit being ambushed is to charge into and through the ambush. Charging the ambush may not sound intuitive but it is a better option than remaining in a planned kill zone or hastily retreating into what is probably a booby-trapped lane of retreat.

Cover Shoot Technique: The *cover shoot* is a technique used to fire on an enemy in positions of cover and concealment. When personnel are being fired upon by a concealed enemy, they may consider following steps (so long as these fit with their standard operating procedures, of course):

- The team splits the target area into zones to ensure interlocking arcs of fire.
- Each team member engages positions of likely cover and places one aimed shot into a box 12" x 12" at *ground level* to the *left* of the cover. After a round is fired, each person should wait for about a second and watch for movement. If there is no movement, then the person should place another aimed shot at *ground level* in a 12" x 12" box at the *right* side of the cover. If there is movement, then the team member is in position to immediately fire another shot into the same box.
- If there is still no movement after the second shot, that team member should seek another position of likely cover further back in his/her zone until the whole area is covered. Once a team member has accurately engaged all positions of likely cover within his/her zone he/she should cease fire and await orders from the team leader.
- The goal of the cover shoot is not necessarily to hit the enemy; instead, firing at the ground near the enemy will kick-up debris, causing the enemy to reveal his/her position and limiting the quarry's ability to effectively return fire.
- The team must maintain an even cadence of fire to ensure a continual "wall" of effective fire is sent down range.
- Note: Under *no* circumstances should the cover shoot be used if hostages, civilians, or military might be injured.

REMEMBER!

Every tactical situation is different from any other. Tactical decisions can only be made by the commander on the ground. Consideration must be given to the effective range of firearms carried, the immediate terrain, and vegetation limitations.

Discussion Questions:

1. Given your Service or Agency's tactics, techniques, and procedures, what actions should your team take if they see the quarry? What if your team and the quarry see each other simultaneously? What if the quarry ambushes your team?
2. What does the signal for "I see the quarry" look like? How will the rest of the team be aware of one team member making this signal?
3. What pre-event indicators might be present if the quarry set up an ambush? What sort of ambush techniques might the quarry in your operating environment use?
4. Remember Cooper's Color Code? What color code state should tracking team member maintain? Why?

8013: Tracking Mission Scenarios (Activity)

Advanced
Practical application
About 10 hours

Terminal Objective: This advanced module gives trainees the opportunity to practice their combat tracking knowledge, skills, and attitudes. Upon completion of this module, trainees should have synthesized their combat tracking abilities, so that they can operationally employ combat tracking. During these exercises, trainees should demonstrate effective application of combat tracking principles, team techniques, and engagement tactics.

Instructor Activities:

1. Preplan the routes and scenarios
2. Prepare any handouts (e.g., litter, maps) that will be given to trainees
3. Divide trainees into their 5–6 member tactical teams
4. Brief each team on their scenarios
5. Allow each team to conduct their follow-up
6. Conduct an after-action review with each team

Synopsis of Module Topics:

Introduction: In this exercise, teams will have the opportunity to practice their active tracking skills, as well as their anti-tracking tactics. Each team will take one turn playing the quarry and one turn playing the tracking team.

Preparation: Instructors must preplan the routes that each "quarry" team will follow. The paths must be carefully selected for difficulty, time required to complete the route, and end locations. If working with multiple teams, ensure routes are planned so that teams do not bump into each other. Also, a central "watering hole" and pick-up locations should be carefully pre-planned. Draw out the quarry routes on map or using grid coordinates. Design a scenario for each quarry team. The scenario should involve a realistic narrative that is relevant to the trainees' operational environment. The scenario should also include specific activities that allow the tracking team to practice their skills, including:

- Micro-tracking techniques
- Getting into the mind of the quarry
- Lost Spoor Procedures
- Enhanced observation
- Concealment and noise discipline
- Encounter action drills

Similarly, think about the skills that the other team can practice. While playing the quarry, that team will have the opportunity to truly "get into the mind of the quarry."

- What are the quarry's motivations?
- What habitual areas, anchor points, or short cuts would they use?
- Would they suspect that they were being followed?
- Would they attempt to employ anti-tracking?
- Would they attempt to set-up an ambush?
- What materials would they intentionally or accidently drop?

You will likely require a convoy that includes a bus, range-approved vehicles, and first aid and water support

Transportation: Trainees and instructional staff should be bussed to their respective follow-up locations. Two teams should be dropped off at each spot, along with two instructional staff members. The vehicles should wait at the watering hole, first, and then later move to the final pick-up location. These key locations should be placed near the end-points of the first and second follow-ups, respectively.

Team Briefings: After trainees are divided into the 5–6 member tactical teams, each team should be briefed. First, the quarry teams should be separated and briefed on their scenarios. Ask them to truly consider their roles. How would the quarry really think and act? Encourage them to give proper cues to the tracking team who will be following them. Also, part of this exercise should entail intelligence collection; thus, the quarry team should have litter, equipment, or other assets that allow them to leave intelligence clues for the follow-up tracking team. After the quarry team understands its goals, they should begin their "escape" from the area.

Next, brief the tracking team. Explain the scenario to the tracking team, and give them whatever typical intelligence they would normally have if this were a real operation. Remind the tracking team about their goals, as well as key combat tracking techniques, such as rapidly closing the time/distance gap, exercising good team formations and communication discipline, maintaining a high degree of security, and constantly seeking relevant information in the environment. Also, ensure the team conducts proper communications with their TOC (which can be played by one of the instructional staff members). Also, remind the team of the early CODIAC lessons on situational awareness, sensemaking, and decision-making. Overall, this briefing should take about 30 minutes (in other words, give the quarry team a 30-minute head start).

Follow-Up: At least one member of the instructional staff should accompany each tactical team (both the trackers and the quarry) to observe their performance. Use the tracking Behavioral Observation Checklist, found on the Resource DVD, to assist with this performance assessment.

After-Action Review: After completing the first follow-up, each team should engage in a debrief with a member of the instructional staff. The follow-up team should describe the operational picture that they built, based upon the intelligence they gathered during the follow-up. Is their narrative accurate? Following the debrief, both the quarry team and the follow-up team should break for lunch at the watering hole. Then the teams should switch roles, and the team briefing and follow-up should be conducted again. Naturally, each team should have different scenarios, mission narratives, and relevant clues to leave behind.

Unit 9: CODIAC Practical applications

This unit gives trainees the opportunity to integrate and apply their CODIAC knowledge, skills, and attitudes in realistic practical scenarios. This unit is the specific implementation of CODIAC skills in the trainees' unique operational domain. Unlike the other modules, instructors will need to invest substantial resources into developing specific content for this unit. A framework and examples are provided; instructors will need to create scenarios appropriately tailored to their trainees.

Suggested Prerequisites:

- CODIAC Units 1–8
- Working knowledge of how to use one's organic optics

Terminal Learning Objective:

- At the completion of this unit, trainees will be able to demonstrate integrate and applied CODIAC knowledge, skills, and attitudes.

Enabling Learning Objectives:

- Use of enhanced observation techniques
- Identification of critical event indicators
- Interpretation of human behavior cues
- Synthesis of ambiguous information
- Proactive analysis and dynamic decision-making
- Employment of cognitive discipline

Estimated Time Allotted for Instruction: About 32 hours

Speciality Support Requirements:

- This unit requires extensive facilitates and staff support. Briefly, successful execution of this unit requires access to one or two military ranges with geographically realistic building layouts, natural or man-made observation points for the trainees, approximately communication networks, a TOC location, 30–60 role-players, and special effects.
- Instructors will also need to invest time creating trainee-specific scenarios. The following pages provide scenario examples, and the appendix includes instructions for range set-up, role-player training, scenario authoring, and other logistical activities.

Example Ville and Scenarios

This section contains example scenarios, developed by Greg Williams and his combat profiling instructional team. These scenarios focus on a generic Middle Eastern town, called Zam-Zam village; they were designed to support warfighter training circa 2010. CODIAC instructors will need to modify these scenarios, or create entirely new scenarios, to support their trainees, the trainees' specific missions, and the nature of the current operational environment. Thus, these scenarios are provided as examples, not as complete scripts.

Overhead view of the "Zam-Zam village," Range Golf, Ft. Bliss, Texas (April 2010)

Ville Set-Up

The following example scenarios take place in "Zam-Zam village," a generic Middle Eastern town, created at Ft. Bliss, Texas. Zam-Zam village is home to a good Sheik, Prince Habibi, and a peaceful Bedouin enclave. However, the town is also inhabited by a "bad Sheik," who seeks to exploit the instability of the region to gain personal fortune and power.

The ville is designed with a high degree of cognitive realism. In other words, although the "houses" are made of CONEX boxes (i.e., freight containers), trainees can readily identify what each building or objects signifies. Additionally, subtle details, such as graffiti, are included to complete the environment.

The range is laid-out differently than a traditional kinetic range. Instead of closely packed buildings, the structures are spread-out and placed in a more realistic manner. The accurate placement of buildings is critical; a standard room-clearing range will not support effective CODIAC scenarios.

Outside of the immediate village, four Observation Points (OPs) were created. Since the terrain at Ft. Bliss is relatively flat, CONEX boxes were used to create man-made OPs. The OPs were established at various distances, on different sides of the village. The following distances were measured from each OP to the center of the village, where the "well" of Zam-Zam is found:

- OP-1: 1000m
- OP-2: 840m
- OP-3: 600m
- OP-4: 480m
- TOC: Next to OP-3

The TOC was located next to OP-3. TOC staff worked in a military-style tent and were effectively "blind" to the village.

↘ RESOURCE:
Resource DVD, 9000.1 – Williams Talks about the Original Range Video

↘ RESOURCE:
Resource DVD, 9000.2 – Range Overview, Updated Range Video

Top Row: The "Bad Sheik's" area included graffiti, a drug lab (far left image), and an HME facility

Second Row from the Top: The marketplace

Third Row from the Top: Village public areas

Left: The village police's vehicle checkpoint

Center: Well of Zam-Zam

Right: The Tire Shop

Bottom Row: OP-3 and the TOC

Scenarios

This POI lists 10 scenarios, but instructors are encouraged to adjust that number to best fit their specific training goals. Enough scenarios should be offered so that all trainees have the opportunity to operate from most, if not all, OPs and spend at least one turn in the TOC.

Scenario Sets

Each scenario should give the trainees opportunity to practice the specific skills listed in each of the Unit 9 modules. Additionally, the scenarios should build upon one another, collectively telling a story. Like all good stories, the overall scenario set should increase in intensity over time, have a dramatic climax, and then reach resolution by the Final Exercise (FinEx).

Scenario Timing

Each scenario should run approximately one hour. The first 15 minutes of each scenario are designed to give trainees time to develop their baselines. Thus, no anomalous activities should occur in the first 15 minutes.

Trainee Placement

Trainees should be divided into the 5–6 person teams. Each team should be assigned to an OP or the TOC. If too few OPs are available, two teams can be assigned to a single OP, but this is not optimal. At least one member of the instructional staff must be embedded with each team. This staff member will deliver the After-Action Review (AAR) for that team.

AARs

Following each scenario, the instructors must deliver individual AARs to each team. The AAR is critical, and a substantial portion of the training will take place during this reflective period. Each AAR may take around 30 minutes to an hour, depending upon the trainees' performance.

Trainee Rotation

After completing a scenario's AAR, trainee teams should rotate to their next position (i.e., either the next OP or the TOC). This down-time also gives trainees the opportunity to rest and discuss tactics amongst their team; so, the transition time should not be rushed. Give trainees approximately one hour to rotate to their new positions. The video listed below gives instructors an example of these instructions, as they were briefed to trainees during the 2010 Border Hunter exercise.

RESOURCE:
Resource DVD, 9000.3 – General Instructions to Trainees Video

TOC

Each trainee team must be stationed in the TOC for at least one scenario. However, do *not* post trainees in the TOC during the FinEx. Past experience has shown that trainees assigned to the TOC during the FinEx (which is the climax of the scenario narrative) are disappointed and frustrated.

TOC Activities

The team posted in the TOC is responsible for managing communications between teams and with outside entities (such as role-played village police, other military units, or distant law enforcement sectors). For instance, if the trainees identify an IED attack left-of-bang, then it will be the TOC's job to warn the (role-played) personnel who are entering the village.

The TOC is also responsible for maintaining the *common tactical picture*. An accurate common tactical picture is a measure of the trainees' collective success; however, if the common tactical picture is inaccurate, then the trainees' observation, orientation, decision-making, and/or communication skills need additional practice.

The video listed below gives instructors a brief glimpse into the TOC at the 2010 Border Hunter exercise.

RESOURCE:
Resource DVD, 9000.4 – Tactical Operations Center Example Video

TOC Equipment

Notice that the TOC must be equipped with communications gear, maps, and other note-taking materials (such as white boards, notebooks, and reams of large paper). Refer to the TOC overview in Unit 7 for additional ideas of the charts/visual aids that might be useful to the TOC.

A role-player poses as an insurgent during the Border Hunter scenarios

MSEL

Each scenario must be crafted with a Master Sequence of Events List (MSEL) script. A MSEL (pronounced like "measle") is a "collection of pre-scripted events intended to guide an exercise toward specific outcomes" (*Joint Master Scenario Event List V2.5 User Guide*, 2006). In other words, a MSEL is the list of key events that occur in a scenario. Each event should be associated with a particular learning objective, as well as performance criteria.

More information on creating MSEL scripts is available in the appendix. Also, the video listed below gives instructors a glimpse at how the role-players may be briefed on the MSEL script.

Role-Players

CODIAC scenarios require a substantial number of role-players, who play the "bad people," "good people," and fill-out the town. Some role-players simply carry out the roles of shopkeepers or average citizens. These "normal" roles are critical, since they help trainees establish the baseline for an area. Without such a context, trainees will be unable to accurately employ their CODIAC skills.

➘ RESOURCE:
Resource DVD, 9000.5 – General Instructions to Role-Players Video

Scenario Controllers

At least 8–10 instructions staff members will be required to execute the scenarios. Additionally, added assistants will make the scenarios run more smoothly. At a minimum, each OP and the TOC will require a staff member, and a "puppet-master" will be needed to orchestrate the scenario events. Other staff members, dressed in specific clothes that identify them as outside the action may be required to act as runners, within the scenario action. For instance, in the example videos, viewers may occasionally notice a person wearing bright red; this individual is out-of-place and is part of the instructional staff, providing instructions to the role-players.

9001: "Pulling Teeth"

Intermediate
Practical application
About 3 hours

Scenario Objective: This scenario gives trainees the opportunity, probably for the first time, to apply their CODIAC observation and decision-making skills in a realistic setting.

Example Videos on DVD:

- 9001.1 – AAR, Video
- 9001.2 – AAR, Part 2, Video

KEY SKILL

Making innovative use of optics to help construct a baseline or profile

This first scenario emphasizes observation, orientation, and the use of intra-team communication to establish a baseline. Before the scenario begins, instructors should reiterate some of the important observation lessons, such as encouraging trainees to use their optics in innovative ways, encouraging open discussion of what trainees observe, and managing the natural limitations of human observation. During the AAR, instructors should discuss the quality of the observation techniques used and ask trainees to articulate the baselines they perceived.

Example Scenario:

This scenario involves a sophisticated ambush by the insurgents. A single sniper shoots a known authority within the ville. Ultimately, the attack will seem like a shell game to the trainees. They will have to decide: where is the attack coming from; who is the intended target; and who actually carried-out the shooting?

Synopsis of Scenario:

Introduction: The first CODIAC scenario is dubbed "pulling teeth," because it may feel slow and obvious to the instructors. Trainees, however, will likely feel overwhelmed by the scenario's stimuli, and they will struggle to identify those key anomalies that forewarn a critical event. Even though this is the simplest scenario, there will be many cues for the trainees to watch. The trainees are expected to see events right-of-bang, but they should start to realize what indicators they missed during the after-action review.

Preparation: For this scenario, and all of the other CODIAC scenarios, instructors will need to prepare their scenario kit, including a MSEL script, required role-player list, required props list, and instructional staff support duties. More information on these preparation requirements is listed in the appendix.

Areas of Instruction: Scenario #1 should give trainees opportunity to practice their observation, orientation, and communication skills. In particular, instructors should emphasize the use of optics and observation to develop a baseline of the village and its inhabitants.

After-Action Review:

- Did teams use all of their organic optics?
- Did teams attempt to make innovative use of their optics, such as using thermals in daylight to search for body bombers?
- Did any team member experience focus lock, change blindness, or any of the other natural limitations of observation? If so, did that team member or his/her teammates solve the problem?
- Did teams use *interlocking lines* of observation? Did they assign different team members to different optical devices (including some using their naked eyes)?
- Did teams engage in constant intra-team communication? Did they discuss their observations, engage in team sensemaking, and create a shared operational narrative? Did they establish a shared baseline?

9002: "Naming Names"

Example Scenario:

This scenario occurs the day after Scenario #1 (in the narrative). The ville has asked the Red Crescent to provide aid following the previous day's attack. As the Red Crescent enter the ville, insurgents stage an ambush on their caravan (seeing it as a target of opportunity). As the insurgents enter, the trainees will have their first opportunity to really see their adversaries.

Synopsis of Scenario:

Introduction: This scenario is called "naming names" since trainees should be actively identifying key people, habitual areas, and anchor points during it. They should begin to see the players in the village, their key locations, and the relationships among the village's inhabitants. Still, as before, trainees are expected to see events right-of-bang.

Areas of Instruction: Scenario #2 should builds upon the lessons of Scenario #1. Trainees should now feel (somewhat) comfortable observing the village from a distant OP. They should continue to make effective use of their optics, observation abilities, and team sensemaking. In addition, they should begin to establish shared operational narratives of the village. They should be able to identify the "good" civilian leaders (both formal and informal), the "bad" insurgent leaders, the village's habitual areas, and the insurgents' anchor points. Additionally, they should be able to explain their reasoning; in other words, if trainees identify a particular villager as the insurgent leader, they should be able to "prove it" by articulating the cues that led to their decision.

After-Action Review:

- Did trainees identify (and tell the TOC about) the *anchor points* and *habitual areas* they observed within the ville?
- Did trainees give these key locations memorable nicknames?
- Did they get a sense of the actives that occur at these locations and the people who use them?
- Did trainees identify (and tell the TOC about) key leaders, hostile people, or other High Value Individuals (HVIs) within the ville?
- Did the trainees use these HVIs' real names or assign them memorable nicknames?
- How would they describe these people? Did they get a sense for their typical behaviors? Their personalities? Their motivations?
- Did trainees see any anomalies? Did they communicate these anomalies amongst themselves? To the TOC?

Intermediate
Practical application
About 3 hours

Scenario Objective: This scenario focuses on identifying key locations and people within the ville.

Example Videos on DVD:

- 9002.1 – Scenario 02, Video
- 9002.2 – Scenario 02, Video

KEY SKILLS

Effectively and efficiently identifying leaders

Looking for signature locations (e.g., habitual areas) through a cluster of cues

This scenario emphasizes identification of key people and places. Before the scenario begins, instructors should reiterate the importance of identifying leaders, habitual areas, and anchor points. Remind trainees to "prove it," that is, to explain their rationale for identifying a specific person or place. Remind them to also use nicknames and to clearly communicate their raw information *and* analyses appropriately. During the AAR, instructors should discuss the quality of the operational narrative the trainees formed and whether they are beginning to develop a sense of the key players' personalities.

9003: "Building a Story"

Intermediate
Practical application
About 3 hours

Scenario Objective: This scenario focuses on encouraging trainees to collectively build shared operational narratives. They should now be able to predict future events, based upon their knowledge of individuals' and groups' past actions.

Example Videos on DVD:

- 9003.1 – Scenario 03, Video
- 9003.2 – Scenario 03, Video

KEY SKILL

Looking for signature behaviors (e.g., of a high-value target) via a cluster of cues

This scenario emphasizes collective sensemaking, shared situational awareness, and the active creation of shared operational narratives through mental simulation and collective discussion. Before the scenario begins, instructors should reiterate the importance of using "storytelling" to predict future events and generate likely courses-of-action. During the AAR, instructors should discuss the quality of the predictive narratives that trainees developed, as well as the techniques they used to generate these predictions.

Example Scenario:

Again, this scenario occurs the day following Scenario #2 (in the narrative). Red Crescent personnel begin strengthening their security posture, and as a result, the insurgents decide to fight back. The insurgents stage a sophisticated ambush, involving a vehicle-borne sniper, a vehicle-borne IED (VBIED), and another large IED. A significant number of dangerous events occur during this scenario, and the set-up for these events (e.g., constructing the VBIED) occurs during the scenario. However, the trainees are unlikely to pick-up on all of the pre-event indicators, just yet.

Synopsis of Scenario:

Introduction: By this time, trainees should be "getting a feel for" the village and its inhabitants. Trainees should have a sense of key individuals' personalities and be able to use *tactical cunning* to anticipate their reactions to events. For instance, in the example scenario (above), trainees should be able to anticipate that the insurgents will respond violently to the Red Crescent's increased security posture.

Areas of Instruction: Again, building upon the previous lessons-learned, trainees should continue to demonstrate effective observation and orientation skills, and they should also continue to pay particular attention to the key persons and places within the village. Now, they should also be anticipating individuals' reactions, based upon inferential knowledge of their personalities and observation of their past behaviors. Trainees should be actively engaging in mental simulating, attempting to determine the most likely courses-of-actions. Further, they should be collaboratively building shared operational narratives and discussing potential courses-of-action as a team.

After-Action Review:

- Were the trainees able to take on any of the HVIs' perspectives? Were they able to employ tactical cunning?
- Did the trainees collectively engage in what-if simulation or otherwise attempt to predict future outcomes as a team?
- Were the trainees able to anticipate what would happen next, after they observed a cue or anomaly?
- Could they induce a pattern from a few individual cues?
- Did they generate explanatory storylines that tie individual items of information together?

9004: "Left-of-Bang Decisions"

Example Scenario:

This scenario occurs the day after Scenario #3 (in the narrative). Following the previous attacks, the ville's police begin increasing their security efforts, conducting more efficient searches at the ville's Entry/Exit Control Point (ECP). The police arrest an insurgent for possessing clandestine weapons while entering the ECP, and this instigates a chain of events. The insurgency wants the arrested insurgent returned, and in response, they attack the ville. They conduct a filmed execution and engage in other violence as a distraction to draw people's attention as they break their arrested brother out of jail.

Synopsis of Scenario:

Introduction: By this point, trainees should be readily identifying anomalies and predicting critical incidents left-of-bang.

Areas of Instruction: Building on past lessons, the trainees should now be efficiently employing their optics and observation techniques, actively communicating, and communicating meaningful (not just descriptive) information as well as analyses. Teams should be articulating operational narratives amongst themselves, and teams should be communicating timely and relevant information among one other and to the TOC. At this point, teams should be predicting critical incidents early enough that they can communicate a warning to the TOC. Further, trainees should articulate their observation, predictions, and rationale clearly—so that the personnel in the TOC can understand develop a shared operational picture with the teams actively observing the village.

After-Action Review:

- Were trainees able to detecting an unfolding event or activity by identifying a piece of it and inferring the rest?
- Did trainees identify any of the critical events left-of-bang? Did they communicate a warning to the TOC?
- Did trainees use appropriate criteria and sound reasoning when predicting the critical incidents?
- Did trainees continue to employ tactical cunning? Did trainees continue to engage in team sensemaking and discussion of cues and their potential meanings?
- Were trainees able to effectively and efficiently communicate their observation, analyses, and rationales to each other? To the TOC?
- If they were able to forewarn the TOC, did trainees recommend any potential courses-of-action?

Intermediate
Practical application
About 3 hours

Scenario Objective: By this time, trainees should be seeing anomalies and starting to predict events left-of-bang. They should also be developing confidence in their decisions and providing the TOC with early warning about upcoming critical incidents.

Example Videos on DVD:

- 9004.1 – Scenario 04, Video

KEY SKILL

Using appropriate criteria (e.g., three cues) to make timely but accurate decisions

This scenario specifically emphasizes making decisions left-of-bang. Building upon the previous scenarios, this scenario gives trainees the opportunity to integrate and apply their developing CODIAC abilities. Before it begins, instructors should remind trainees to look for anomalies, use appropriate criteria to predict future events, and then act quickly and appropriately upon their predictions (in this case, by communicating with the TOC). During the AAR, instructors should discuss the ability teams to make appropriate decisions left-of-bang

9005: "Running Man"

Intermediate
Practical application
About 3 hours

Scenario Objective: This module gives trainees the opportunity to practice observation at night and get comfortable using their thermal optics.

Example Videos on DVD:

- 9005.1 – Scenario 05, Video

KEY SKILL

Detecting an unfolding event or activity by identifying a piece of it and inferring the rest

Since this is a night scenario, the trainees will have much less visibility. They will need to make more inferences and put more energy towards developing a comprehensive story from the few cues they can observe. Before the scenario begins, instructors should remind trainees that they are able to predict future events from inferring only a few cues. Remind them to trust their judgment and continue sensemaking—even though it may seem more difficult at night.

Example Scenario:

Because this is a night scenario, its narrative timing will be a bit different from the others. Mainly, this is because the trainees will likely perform poorly during the night scenario, and it is important that their poor performance does not affect the overall scenario-set's narrative.

The townspeople have asked the UN (or some similar security force) if they may go out after curfew to mourn the villagers who died during Scenario #1. Essentially, this will be like a massive "block party," held outside the ville. One of the insurgents will show up at the event, and the trainees will have to determine whether this is a good or bad occurrence. In actuality, he is performing a body dump; when identified, he will take off on foot, running from the police. He will disappear into the crowd. The trainees will need to identify the "bad guy" from the crowd. They can do this, in part, by looking at the heat signature of the "running man." Since he ran from police, his body temperature should be higher than that of the crowd in which he is hiding.

Synopsis of Scenario:

Introduction: Similar to Scenario #1, this scenario emphasizes basic observation, orientation, and communication. Although trainees should now be comfortable observing the village during the day, they will likely struggle with night observation.

Areas of Instruction: The primary goal of this scenario is to allow trainees to practice their night observation and optics use. Instructors should reiterate the observation and orientation lessons from Scenario #1, but this time, they should also stress technical proficiency with night/thermal optics. Following the scenario, instructors should lead the trainees in discussions, attempting to predict the ville's behaviors for the next day. At this point, the trainees will have spent roughly a week (in scenario) observing the village.

After-Action Review:

- Going back to the lessons of Scenario #1, did teams use all of their appropriate organic optics?
- Did teams effectively use their night-vision and thermal optics?
- Did any team member experience focus lock, change blindness, or any of the other natural limitations of observation? If so, did that team member or his/her teammates solve the problem?
- Did teams use *interlocking lines* of observation?
- Did any team members have technical issues with their gear?

9006: "Hurry it Up"

Example Scenario:

Because of the recent catastrophes, the ville's leader, Prince Habibi, begins recruiting new support for Zam-Zam's security force. He puts up posters, has meetings, and begins a recruitment drive. Habibi also holds a "mini" *shura* (Islamic community meeting) with the other powerful individuals in the ville. This is slap-in-the-face to the insurgency. They are trying to extort the village—trying to have the ville follow the insurgency in return for peace—but their plans will not work if Habibi fights back. In retaliation the insurgents stage a complex ambush in the Bedouin Village. The trainees should be asking themselves: Will the Bedouins decide to embrace the insurgency or will they hold together? This scenario will be as tough as Scenario #1, but this time, the trainees should be able to identify the indicators left of bang.

Synopsis of Scenario:

Introduction: This scenario continues to build upon the previous scenarios. Trainees should now be demonstrating smooth, effective use of their CODIAC skills. This scenario specifically emphasizes efficiency. Trainees must continue to operate accurately and securely, but are they being efficient? How can they accelerate their OODA-Loops? How can they improve their overall tempo?

Areas of Instruction: This scenario continues to allow trainees the opportunity to practice their core CODIAC skills. In addition to observing, orienting, and making decisions, the trainees should be practicing sound security procedures. Are they using *guardian angels*? Are they providing *overwatch* for one another? Are they using *interlocking lines* of observation, communication, and intelligence?

After-Action Review:

- Were trainees able to efficiently detecting unfolding events by observing just a few cues and inferring the rest?
- Did trainees look for prototypes to guide their rapid recognition and decision-making?
- What techniques did the trainees use (or could have used) to accelerate their OODA decision-making processes?
- Did trainees effectively communicate their observations, analyses, and recommendations to the TOC? Did they do so in a timely manner?
- Were teams mindful about their own security posture? Did teams assign a *guardian angel*? Did teams employ *overwatch*?

Advanced
Practical application
About 3 hours

Scenario Objective: This module continues to give trainees the opportunity to practice their CODIAC skills. Trainees should now be operating more rapidly.

Example Videos on DVD:

- 9006.1 – Scenario 06, Video
- 9006.2 – Assassination of the Bedouin Leader, Video

KEY SKILL

Looking for prototypes to guide rapid recognition and decision-making

This scenario continues to emphasizes collective sense-making, shared situational awareness, creation of shared operational narratives, and rapid decision-making. Before the scenario begins, instructors should reiterate the importance of looking for prototypes to guide efficient recognition and decision-making. During the AAR, instructors should ask trainees what prototypical matches they made and discuss the effectiveness and efficiency of the trainees' decision-making.

9007: "Stay Skeptical"

Advanced
Practical application
About 3 hours

Scenario Objective: Trainees should now feel comfortable with their CODIAC skills. Instructors should take this opportunity to remind them to question everything, being skeptical about what they see and interpret at all times.

Example Videos on DVD:

- 9007.1 – Scenario 07, Video

KEY SKILLS

Taking an evidence-based approach, using hard data to confirm or disconfirm hypotheses

Not settling for unexplained events or evidence but looking for antecedents to a situation

This scenario emphasizes taking an evidence-based approach, looking for hard data to explain what is really seen (rather than relying upon potentially biased assumptions), as well as actively searching for additional hard evidence when something unexpected occurs.

Example Scenario:

The ville is now much more secure. They hold regular patrols, and the insurgents now find it difficult to bypass the security. The ville has finally become a "hard target." Taking advantage of his new position of strength, Prince Habibi tries to make to make peace with the insurgents. First, he sends a messenger to discuss a cease-fire with the insurgents, but they beat the man and send him back to the Prince. Next, the Prince tries meeting with the insurgents, personally. He holds a *shura* with the leader of the insurgents, and they agree to a cease-fire. This appears to be a good omen, but can the insurgents be trusted?

Synopsis of Scenario:

Introduction: As before, the trainees should be developing their applied skills; they should be using their optics and observation skills effectively, making sense of the cues they see, developing operational narratives, and communicating more efficiently. As the trainees continue to practice these skills, they should be reminded not to become too complacent.

Areas of Instruction: This scenario emphasizes the need to be adaptable, keeping in mind that unexpected events may always occur. What do the trainees do when something unexpected takes place? How do they reconcile their operational narratives? Do they look for new evidence and update their collective "stories," or do they turn to their own preconceived notions of what *should* happen?

While trainees must trust their own abilities and insights, they must also learn to question everything. Trainees must take an evidence-based approach, always asking themselves and their teammates to "prove it." In other words, when one of the trainees makes an inference, he/she should be able to clearly articulate the rationale of that decision. Similarly, trainees should actively search for additional information; they should look for reasons why certain events occurred and search for evidence to support their assumptions.

After-Action Review:

- Did the trainees develop shared operational narratives of the scene? Were they able to predict what would happen?
- Did they search for evidence to explain unexpected events? Were they able to generate storylines of why these unexpected events occurred?
- Did they take an evidence-based approach, using hard data to confirm or disconfirm their hypotheses?

9008: "Just Ask"

Example Scenario:

The trainees have the opportunity to meet with leader of the ville. Each trainee team will have 20 minutes to talk face-to-face with the Sheik. They can ask as many questions as they want, but they should use the tactical questioning and interview techniques taught in class. If the trainees perform well, then they will gain useful intelligence, but if they perform poorly, they may upset the Sheik, causing him to leave. While one team is interacting with the Sheik, the other teams can rest and talk.

Synopsis of Scenario:

Introduction: This scenario is different from the others. Scenario #8 emphasizes close-in, HUMINT intelligence collection.

Areas of Instruction: Trainees must use their knowledge of the village, their observation skills, and their understanding of human behavior to glean intelligence from the village leader and gain access to the village.

In addition to practicing their HUMINT related skills, the trainees should also be thinking about observing the villagers' biometric, kinesic, proxemic, and atmospheric cues. Can they effectively (and subtly) observe the villagers while also carrying on a discussion? Plus, trainees (as always) must remain conscious of their own security posture. Are they using the five combat multipliers? Are they maintaining overwatch? Do they have a plan for egress, if the meeting becomes compromised?

After-Action Review:

- Did teams employ tactical cunning? Did they attempt to get into the minds of the insurgents, as well as the villagers?
- Did teams attempt to be *good shepherds*, attempting to win the "hearts and minds" of the civilians?
- Did the teams notice any places where the village leader saved (or lost) face? What did the trainees do (or could have done) to help the leader save face?
- What biometric, kinesic, proxemic, and atmospheric cues did the trainees observe?
- Did the trainees discuss cues amongst themselves during the scenario? Did they engage in open discussion or were they more subtle?
- Did the trainees use appropriate cultural references?
- Were teams mindful about their own security posture? Did teams assign a *guardian angel*? Did teams employ *overwatch*?

Intermediate
Practical application
About 3 hours

Scenario Objective: This scenario gives trainees the opportunity to practice their HUMINT intelligence collection and tactical questioning skills.

Example Videos on DVD:

- 9008.1 – Scenario 08, Video

KEY SKILL

Effortlessly using observation techniques that do not require conscious attention

This scenario emphasizes HUMINT collection. As part of the intelligence collection process, trainees must engage in observation while simultaneously carrying on a conversation with the village leader and maintaining their own security posture. Before the scenario begins, instructors should emphasize the importance of carrying out all of these critical tasks simultaneously. Discuss ways to distribute the tasking amongst the team, and discuss the importance of pre-planning roles and responsibilities.

9009: "Keeping Cool"

Advanced
Practical application
About 3 hours

Scenario Objective: This scenario is designed to stress trainees, giving them the opportunity to practice their CODIAC skills while under psychological duress and a sense of time pressure.

Example Videos on DVD:

- 9009.1 – Scenario 09, Video

KEY SKILL

Employing stress reduction strategies to manage physiological stress reactions

This scenario emphasizes effective decision-making despite the presence of stress and psychological duress. Before the scenario begins, instructors should remind trainees how to use effective stress reduction strategies (deep breathing, concentrating on task at hand, pausing to reflect) and to back-up their fellow teammates if they show signs of overload.

Example Scenario:

After meeting with the Sheik (in Scenario #8), the trainees have gained some access to the ville; they are now able to act, at least a little. During this scenario, the trainees have been asked (e.g., by the local police or an international force) to provide overwatch of a route while one of their teams (comprised of role-players) clears it. The route clearance team will make a series of mistakes, leading up to a critical failure. The trainees should anticipate the danger and warn the route clearance team. However, even with the warning, the route clearing team will continue to make errors, and the errors will lead to a catastrophic outcome. In this case, the team is ambushed by the insurgents, who fire upon them with an RPG.

Synopsis of Scenario:

Introduction: This scenario is intended to be frustrating for trainees. They will be able to see the eminent danger, but despite their best efforts, they will not be able to convince the role-players to avoid the catastrophe. This scenario should be intentionally designed to stress the trainees.

Areas of Instruction: The goal of this scenario is to give trainees the opportunity to practice their CODIAC skills under stress. Additionally, they should be practicing their stress reduction strategies and providing back-up behavior to one another.

At the end of the scenario, the trainees should be informed that they now have unlimited access to the village; the villagers are ready to accept their help in eliminating the insurgency once-and-for-all. Following the exercise, the trainees will be released to plan for the FinEx. Trainees should be encouraged to discuss their tactics and determine the optimal approach to confronting the remaining insurgents.

After-Action Review:

- Did trainees make effective decisions despite the high stress conditions?
- Did any of the trainees experience tunnel vision, focus lock, or any of the other natural limitations caused by stress and fatigue?
- What stress reduction techniques did trainees use?
- Did any of the trainees exhibit signs of overload? If so, what did those behaviors look like? What did his/her teammates do (or what could they have done) to help?
- Did the trainees employ *overwatch* and assign a *guardian angel*? Was that necessary and/or effective in this instance?

9010: "Final Exercise"

Example Scenario:

This scenario marks the first time that the trainees enter the ville. The trainees occupy two OPs, and then they push ground elements into the ville. The trainees at the OPs will have to direct the ground elements. They have to coordinate movements without anyone being compromised by a sniper, IED, or other attack.

There will be a number of surprise problems that the trainees have to overcome. The trainees will have to conduct legal, moral, and ethical decision-making. They will have to do shooting packages, explaining why and how to kill the various targets. At the end of the exercise, the weapons cache will explode, then the trackers will lay spoor (four trails) that the trainee teams can follow. Each of these will lead to a different, relevant outcome, such as an insurgent who commits suicide, an insurgent who is captured, and so on.

Intermediate
Practical application
About 6 hours

Scenario Objective: The FinEx does not include a single instructional focus, instead the FinEx should give trainees the opportunity to practice all of their CODIAC skills in a single, large-scale exercise.

KEY SKILL

Trusting that one's skills will overcome obstacles in the difficult situations

Instructors should remind trainees to trust their skills. They should all be comfortable with the CODIAC skills; now, they simply have to trust that these skills will be effective.

Synopsis of Scenario:

Introduction: This scenario must be highly personalized to the trainees' organization and operational environment. However, regardless of its specific nuances, the FinEx should provide trainees with the opportunity to practice all of their CODIAC skills.

Instructors are encourage to use "tactical freezes" (essentially a scenario pauses) to control the pace and direction of the FinEx. For instance, each time trainees make a critical decision (e.g., to send a fire team into the lead insurgent's house), a tactical freeze should be called. Then trainees must explain their rationale to the instructors, who will allow or disallow each action. This way, trainees receive immediate feedback and the FinEx is not derailed by a single bad decision.

After-Action Review:

The final scenario's after-action review should encompass all of the CODIAC skill areas:

- Did trainees effectively employ visual observation techniques and make effective use of optics?
- Did trainees identify critical event indicators left-of-bang?
- Were trainees able to interpret human behavior cues?
- Did trainees synthesize the information effectively, collectively anticipating what would happen next?
- Did the trainees proactively analyze the context and make rapid decisions about what they saw?
- Did the trainees employ cognitive discipline, backing-up each other, employing tactical patience, and functioning under stress?

Appendix

Vocabulary

Action Indicators: Foot or body marks left upon the ground (such as a person sitting down to change his/her shoes) that indicate that a certain identifiable action has taken place.

Active Track: Actively following a set of tracks while the quarry is still moving.

Adoration: When used as a precise keyword, adoration refers to subordinates' behaviors that show reverence and submissiveness to their leader. It is associated with the *proxemics* domain.

Adrenaline: A hormone released in the body during fight–flight–freeze conditions.

Aerial Spoor: Damage and disturbance to vegetation created by the quarry, found from foot to head height.

Alleyway Scans: These scans are used in tracking operations when the spoor follows an existing track, trail, riding path, or stream bed. It is effective when the spoor is obvious and can be seen from a distance.

Anchor Point: An area where only certain individuals frequent without reservation; individuals outside of the permitted group or sect have reservations about entering such areas.

Anomaly: An anomaly is the presence, absence, or change of something that creates a deviation from the baseline.

Anti-Tracking: Techniques used by a quarry to disguise or conceal its spoor and attempt to fool the tracker.

Atmospheric Shift: A sudden change to the "feel" of an area, usually indicating danger. Atmospheric shifts are associated with the *atmospherics* domain.

Atmospherics: One of the six domains of combat profiling, atmospherics are concerned with interpretation of environmental mood of an area, including the look, sound, taste, smell, and feel of a location.

Automaticity: This is the learning of a task to a point that it becomes essentially attention-free. This is why we practice gun-drills and immediate action drills repetitively, so we do not have to think about them under stress.

Average Pace Method: A quarry estimation technique in which a line is drawn behind one heel print (key quarry print) and another behind the next, opposite foot of the same individual. The total number of prints inside the two lines are then counted, and the total is divided by two. This number will result in an approximate number of quarry. This method is effective for quarries consisting of up to 15 people.

Back Tracking: Following a track backwards, from the quarry to the track's origin. Back tracking supports intelligence collection.

Balance: Used in tracking to indicate a state of physical equilibrium, with weight equally distributed on the right and left foot.

Baseline: An initial set of critical observations, or data, used to establish the norm of person or place. Baselines are dynamic and will continually evolve.

Binocular Vision: Binocular vision occurs when an object is viewed with both eyes; objects are perceived in three-dimensions.

Biometrics: One of the six domains of combat profiling, biometrics are concerned with the interpretation of physiological reactions which are autonomic instinctive unlearned reaction to a stimulus.

Bitterman Technique: A tracking technique that is used when the quarry follows a specific azimuth or line of travel. The tracking team can determine this azimuth by aligning team members along the track line and "shooting" an azimuth using a compass.

Blushing: Developing a ruddy appearance, or red face, due to embarrassment, shame, or emotional upset. Blushing is a *biometric* signal.

Bottom-Up Processing: Perceiving stimuli through the sensory systems. In contrast to *top-down processing*, bottom-up processing is not affected by the brain's preconceived expectations.

Box Search: A combat tracking *lost spoor procedure* in which natural lines surrounding an area (such as a fence) are identified and systematically searched for evidence of tracks cutting that line. If no tracks are found along the box's perimeter, then the quarry is likely still within the box.

Cerebral Cortex: An evolved part of the human brain, responsible for conscious experience, perception, thought, and planning.

Change Blindness: Humans are blind to change when their attention is focused, either visually or mentally (for example through top-down processing).

Channel: In combat profiling, a channel is an environment features that funnels or guides people's movement through an area. Channeling is associated with the *geographics* domain.

Channel Capacity: The maximum data rate that can be attained or maintained by the brain; channel capacity is typically 7±2, unless under stress, when it drops to around 3.

Chunk: A mental grouping; experts appear to process more information at a time, because they use top-down processing and mental file-folders to chunk (cluster) bits of information together.

CLIC: In the Marine Corps, the Company Level Intelligence Cell (CLIC) give a small-unit capacity to develop actionable intelligence.

CLOC: In the Marine Corps, the Company Level Operations Cells (CLOC) is the location where information is aggregated to provide situational awareness for the commander; a small-unit Tactical Operations Center.

CODIAC: Combat Observation and Decision-making in Irregular and Ambiguous Conflicts—this instructional set.

Cognitive Illusion: Cognitive illusions occur when the brain makes (incorrect) unconscious inferences. In general, they can only be overcome through experience and training.

Cognitive Load: The load on the information-processing system, especially working memory. Since working memory is limited by size and duration, humans can only processes a certain amount of information at a given time.

Collection (Intelligence Cycle): During this phase of the intelligence cycle, personnel collect information and communicate it to the command element.

Color Change: In combat tracking, a subtle color change on the ground, caused by ground disturbance.

Combat Multiplier: A supporting means that significantly increase the relative power of a force while actual force ratios remain constant.

Combat Profiling: The art of identifying behavioral cues, synthesizing them into a meaningful pattern, and then making sense of that pattern, ideally, left-of-bang.

Combat Tracking: The art of following the spoor and sign left by a quarry while in a dangerous area or adversarial position.

Common Tactical Picture: An accurate and complete display of relevant tactical data that integrates tactical information from the multi-tactical data link network, ground network, intelligence network, and sensor networks.

Comparison Method: A quarry estimation technique where the tracking team compares their own tracks in a single-file line to those of their quarry.

Conclusive Evidence: In tracking, tracks or other evidence left on the ground that are indisputably left by the quarry.

Cone Cells: Eye cells located in the central portion of the retina, which are used for day vision, distinguishing color, and sharp contrast.

Contamination: In tracking, tracks or other disturbances made by anyone or anything, other than the quarry, that obscures or obliterates the quarry's spoor.

Context and Relevance: The background, setting, or situation surrounding an event, and the meaning or importance of something in relation to the context.

Cooper's Color Code: Cooper's Color Code is a system for describing the levels of awareness.

Cornea: The clear covering over the pupil; this portion of the eye bends most of the light rays to focus and it ensures that nothing enters the pupil.

Counter-Tracking: Measures employed to harm pursuing trackers or psychologically damage the tracking team.

Counterinsurgency (COIN): Comprehensive civilian and military efforts taken to simultaneously defeat and contain insurgency and address its core grievances.

Cover Shoot: A combat tracking *encounter action drill* used when the tracking team is fired on by the enemy from positions of cover and concealment.

Cross-Over: A *lost spoor procedure* in which the two flanker "cross over" to the other side of the formation while searching the ground for spoor.

Demographic: An overt, population characteristic of a person, such as race, age, income, or educational attainment.

Desert Shine: Color change that occurs when the quarry flattens vegetation or a ground surface, making it appear smooth and, consequently, shiny.

Detailed Search: A visual search technique using the *overlapping strip method* to carefully scan from near to far.

Diopter Sight: The diopter is an aperture used to assist the aiming of guns/devices.

Direct Count Method: A quarry estimation technique where the tracking team physically identifies each distinct print along the track line.

Direction: In combat tracking, the quarry's direction of travel. Part of the LiNDATA SitRep.

Dissemination (Intelligence Cycle): During this phase of the intelligence cycle, personnel collect information and communicate it to the command element.

Distributed Operations: Distributed operations describe an operational approach that creates an advantage over an adversary through the deliberate use of separate, coordinated, and interdependent actions.

Disturbance: In combat tracking, disturbance refers to ground that has been moved from its natural state

Divided Attention: Dividing one's attentional processing between more than one task.

Dwell Time: In tracking, the amount of time the foot is on the ground in the same spot.

Emotion–Memory Link: See *Memory–Emotion Link.*

Encounter Action Drill: A combat tracking term synonym for "immediate action drills." Tactical drills designed for swift reaction to enemy contact.

Endorphins: Naturally occurring opium-like chemicals in the brain and nervous system that are released to relieve pain.

Entourage: One or more people following (i.e., in a beta position to) another. Entourage is a *proxemics* cue.

Explicit Knowledge: Explicit knowledge can be written down, transmitted, and understood by others—basic facts and formulas.

Extended Line Formation: A tracking team formation in which the team lines-up perpendicular to the track line.

Fight, Flight, or Freeze: A natural state that occurs individuals are faced with extreme stress. The brain's limbic system takes control from the cerebral cortex, and a person will choose one of three options: fight, flight or freeze.

Flank Trackers: Two personnel in a tracking team, one positioned on each side and slightly ahead of the tracker. Their functions are to protect the tracker and team leader from ambush, to assist in the search for lost spoor, and undertake close-in recon of "track traps."

Flanker 360: A combat tracking *lost spoor procedure*, in which both flank trackers systematically move around in a circle in an attempt to locate the lost tracks.

Flattening: In tracking, marks, such as by the weight of a foot, that flatten the natural texture of the ground.

Flushing: Developing an extremely ruddy appearance across the face and body. Flushing is a *biometric* signal.

Focus Lock: An observational challenge in which the observer becomes fixated on an object. To prevent this, it is important to maintain peripheral vision.

Focused Attention: Attention directed solely to specific stimulus.

Follow-up: The physical act of following a set of tracks left by a specific quarry.

Foot Roll: The rolling motion made by the foot as the body's weight is moved over the foot.

Fovea: The part of the eye responsible for sharp central vision (also called foveal vision).

Full-Spectrum Operations: Full spectrum operations are the range of operations Joint Forces conduct in war and military operations other than war.

Geographic Profiles: The necessary or preferred landscape features associated with a particular person, group, or type of activity. Geographic profiles are related to the *geographics* domain.

Geographics: One of the six domains of combat profiling, geographics is the study of the physical geography, weather, and human terrain of an area, as well as the interpretation of the relationship between people and their physical surrounding.

Ghost Spoor: A phenomenon that occurs when a tracker starts looking for sign and then start to imagine spoor where there is none.

Good Shepherd: One of the five *combat multipliers*, good shepherds build trusted networks, with local allies, community leaders, local security forces, NGOs and even within their own teams.

Ground Spoor: Marks and impressions of footwear and other body parts or equipment, left on ground surfaces.

Guardian Angel: One of the five *combat multipliers*, These are the alert Marines/Soldiers (at least in buddy teams), placed in a covert position that protect their units--using an ambush mentality, unseen by the enemy, watching over their units.

Habitual Area: An area where most individuals within a given group or sect would frequent without reservations. Habitual areas are related to the *geographics* domain.

Half Y Formation: A tracking team formation, same as the *Y formation* except that one of the flankers is removed and placed behind the tracker or team leader.

Hard Target: A person, unit, or vehicle that is protected against attack. The opposite of a hard target is a *soft target.*

Hasty Search: A visual search technique, used as the first phase of observing a target area. The observer conducts a hasty search (about 10 seconds) for any enemy activity immediately after taking up position.

Heel Strike: The *ground spoor* left by a the heel of a foot stepping on the ground.

Heuristics: One of the six domains of combat profiling, heuristics are rapid methods of mentally imprinting and labeling observed behaviors. They are "tactical shortcuts" for the brain.

Histamines: Natural body chemicals that trigger an inflammatory response. Histamines are related to the *biometrics* domain.

HUMINT: "Human Intelligence," it refers to gathering intelligence through interpersonal contact.

Ideology: A person's world view, ideologies are the ideals, goals, and expectations that guide actions. For this training, an ideology contains three relevant parts: culture, politics, and religion.

Initial Commencement Point: In tracking, the point where a tracking team commences following the spoor. This need not be the site of the incident, but could be at another point somewhere along the trail.

Interlocking Lines: One of the five *combat multipliers*, interlocking lines of fires, observation, and reporting should be employed. Interlocking lines ensure that personnel cover the gaps and seams of their operational area.

Irregular Warfare: A violent struggle among state and non-state actors for legitimacy and influence over the relevant populations.

Key Print: Used in quarry estimation, the key print is one clearly distinguishable footprint among a group of prints.

Kinesics: One of the six domains of combat profiling, kinesics involves interpretation of body movements, facial expressions, and other nonverbal cues.

KOCCOA: An acronym used to remember high priority terrain features: Key terrain features, Observation points, Cover, Concealment, Obstacles, and Avenues of approach.

Last Known Spoor (LKS): In tracking, the location of the last confirmed spoor.

Lecturing: A *kinesics* cue in which a person points his/her index finder and wags his/her hand up-and-down (as if angrily lecturing someone).

Left-of-Bang: Thoughts or actions that occur left-of-bang happen *before* a critical event. Left-of-bang actions are proactive, occurring before the enemy can carry-out his/her violent act.

Limbic System: An "older" part of the brain (in terms of evolution) involved in instinctive behavior and emotions.

LiNDATA: In tracking, the report sent to a command element; LiNDATA stands for Location, Number, Direction, Age, Type, and Additional information.

Litter: In tracking, any man-made artifact that was either accidentally dropped, deliberately discarded, or hidden by the quarry.

Long-Term Memory: The theoretically unlimited information storage center of the brain.

Lost Spoor Procedures (LSP): A systematic set of procedures designed to relocate the spoor when it is lost.

Lugs/Grippers: Deep ridges in and around the center of a shoe's sole that grip the ground. Often found on work boots, such as Vibrams.

Macro-Tracking: A form of tracking in which the tracker looks ahead, searching for the furthest identifiable spoor in order to close the time-distance gap.

Man Tracking: A synonym of *combat tracking*.

Manifesto: A public declaration of an ideology.

Memory–Emotion Link: Associating an emotional response with something that is learned (i.e., a memory).

Mental File-Folder: A set of knowledge and experience about something that is stored in memory. An organized cluster of pre-conceived ideas, associated behaviors, and contextual information. Formally called a *schema*.

Mental Simulation: The process of imagining how one's predictions about a scene may play-out.

Micro-Tracking: A form of tracking in which the tracker carefully examines all spoor and sign, looking for information that will provide tactical advantage or other insights.

Mimicry: When used as a precise keyword, mimicry refers to a person mirroring the body language and/or actions of another. It is associated with the *proxemics* domain.

Mnemonics: Mental tricks that aid memory and retention.

Monocular Vision: Objects seen with only the left or right eye; monocular visions only sees in two dimensions.

Natural Lines of Drift: Most commonly associated with the path of least resistance, natural lines of drift are paths used repeatedly. They become predictable pathways through obstacles.

Natural State: The established, natural state of the ground unaffected by any tracks or sign.

Negative Space: The space between the positive spaces; this is the area of shadow and background activity that an untrained observer often overlooks.

Night Tracking: Nighttime follow-ups, which can be conducted using Infrared/IR lamps, LEDs, incandescent lights, or light sticks.

Nystagmus: Involuntary eye movement, typically caused by ingestion of alcohol or drugs. Nystagmus is a *biometric* signal.

OODA-Loop: A constantly revolving cycle that the mind goes though every second of every day in dealing with all tasks from mundane to the most complicated. The cycle follows the pattern of Observe-Orient-Decide-Act (OODA).

Overlapping Strip Method: The visual search technique used with a *detailed search*. Starting with the area nearest to the observer, the observer systematically searches the terrain, starting at the right flank and then moving toward the left in a 180° arc. Each visual arc includes about 50 meters of depth. After reaching the left flank, the observer searches the next swath nearest to his/her post. Each visual arc overlaps the previous search area by at least 10 meters in order to ensure total visual coverage of the area.

Overwatch: A tactical technique in which one element is located in a position of cover, so as to support another element by providing observation, cover fire, or other security protections.

Pace: The distance covered by a step.

Passive Track: Following a set of tracks, when the tracks are "cold." Normally used for intelligence gathering purposes, such as looking for base campsites or other evidence of insurgent activities.

Perception: The cognitive process by which sensory information is organized and interpreted to produce a meaningful experience of the world. Also, the first level of *situational awareness* is called "perception," and it involves observation, cue detection, and simple recognition of situational elements (objects, events, people, systems, environmental factors) and their current states (locations, conditions, modes, actions).

Perceptual Fill: Rather than perceiving holes in our vision, the human brain "fills-in" portions of the visual scene that are masked by the eye's natural blind spot.

Pitch Angle: The orientation of a foot to the line of travel. A foot can pitch outward, inward ("pigeon toed"), or remain parallel to the line of travel.

Planning and Direction (Intelligence Cycle): During this phase of the intelligence cycle, analysts determine what additional information is needed, and how it should be collected.

Positive Space: Physical terrain features that have mass; solid objects such as buildings, trees, signs, or vehicles. Personnel cannot typically see through positive space, but it naturally attracts the human eye. People are inclined to look from positive space to positive space.

Precipitating Event: An action or activity that brings about a certain outcome; the cause. When identified before a critical event, precipitating events are *pre-event indicators*.

Pre-Event Indicators: An observable cue that suggests a certain future event will occur.

Pressure: In tracking, the total weight of the quarry (to include any load carried), transferred through the foot to exert force onto the ground.

Pre-Terminal Point: The part of the foot that is second-to-last to leave the ground; often the ball of the foot.

Primary Impact Point: In tracking, the first part of the foot to strike the ground.

Processing and Exploitation (Intelligence Cycle): During this phase of the intelligence cycle, the transformation from "information" into "intelligence" begins. Analysts interpret the raw data to identify any useful intelligence.

Production (Intelligence Cycle): During this phase of the intelligence cycle, the significance of the intelligence is established and its implications determined. In this phase, intelligence is combined with data from additional sources, in order to create a robust operational picture.

Prototype: An original form or instance of something that serves as a typical example for items of the same category.

Prototypical Matching: In regard to decision-making, a *prototypical match* is a "close enough" match.

Proxemic Pull/Push: Body language that either draws-in or pushes-away others. A *proxemic push* occurs when a person or group uses body language to create distance to another person. A *proxemic pull* occurs people use body language to invite others toward them. Both are associated with the *proxemics* domain.

Proxemics: One of the six domains of combat profiling, proxemics involves the interpretations of spatial relationships in order determine the dynamics of human interactions. Proxemics is the act of betraying affiliations through the dynamics of proximity.

Psychographic: Psychological features that characterize a person or group.

Pupil Dilation: A physiological response in which the eye's pupil varies in size. It can have a variety of causes, from reaction to light, to narcotics use or observation of an attractive person/item. Pupil dilation is associated with the *biometrics* domain.

Quarry: Used as a synonym for "fugitive," "target," "adversary," or "the pursued."

Quick Scan: The first combat tracking *lost spoor procedure*; the tracker stops behind the last known spoor and quickly, but carefully, scans the ground in a 240° arc in front of him/her. If the quick scan fails, the tracking team can begin using more formal lost spoor techniques.

Rear Security Tracker: The position behind the team leader in a five-person tracking team, who is responsible for rear security, binocular observation, marking the last known spoor, aiding the tracker or flankers, conducting recon tasks, operating the global positioning system, and any other tasks as required.

Regularity: The uniformity and predictability of man-made patterns, such as footprints with a uniform tread.

Retina: The area along the back of the eye that contains two types of light receptors (*rods* and *cones*) for vision.

Rhythm: A stimulus, such as a music beat or footprint, that recurs at regular intervals. Nature has its own rhythm, but usually without regularity. Something in nature that is spaced at regular intervals will stand out as an anomaly from the natural state of the environment.

Right-of-Bang: The time frame following a critical incident. Actions that occur right-of-bang are generally reactive.

Rod Cells: Eye cells located peripheral to the cone cells, which are used for night vision and peripheral vision; they do not see color but are attracted to motion.

Rubble-ing: A damaged, debris-strewn area in which rubble may mask people, paths, or activities (urban masking), as well as create specific movement channels.

Rule of Three: This "rule" reminds personnel that in most cases, a single cue is not enough evidence upon which to make a decision—unless that cue is substantial (e.g., an immediate threat to a person)—however, once three cues (i.e., three anomalies) have been detected, a decision must be made.

Save/Lose Face: Saving and losing "face" are concerned with retaining or loosing the respect (or self-perceived respect) of other people. For instance, if someone is publicly embarrassed, he/she may "lose face." Personnel can help others "save face" by giving them respectable ways to follow requests.

Schema: See *Mental File-Folder.*

Second-Order Effects: The reaction to (or effects of) a first-order effect. The secondary or downstream outcome of an incident.

Selective Attention: See *Focused Attention.*

Sensemaking: A process in which a person or team engages in an effort to understand perceived cues, interpret their relationships, and anticipate the trajectory of a situation. In other words, sensemaking is the ongoing process of giving meaning to one's experiences.

Sensory Systems: The physiological systems used to perceive the world: sight, sound, smell, touch, and taste.,

Sequencing: Sequencing occurs when the brain creates grouping or anticipates a pattern based upon a sequence of observed cues; these perceived groupings or patterns may be accurate or inaccurate. Sequencing usually takes place at the seventh instance of a cue.

Seven-Step Terrorist Planning Cycle: A seven-step process that terrorists follow (not necessarily consciously) when planning, executing, and exploiting their activities.

Sign: The whole group of physical indicators that are not part of *ground* or *aerial spoor*, such as broken cobwebs, disturbed insects nests, *litter*, or water deposits.

Single/Ranger File Formation: A linear staggered formation, similar to the Column Formation, except that the space between personnel laterally is decreased in order to mask the number of persons in the unit.

Site Exploitation: The practice of identifying and processing intelligence at a specific location.

Situational Awareness (SA): An individual's overall understanding of the operational environment, including the time and location of key components, comprehension of their meaning, and a projection of their status in the near future. In other words, SA is internal understanding and integration of the perceived stimuli. It is not a display or the common operational picture; it is the interpretation of displays or the actual observation of a situation.

Soft Target: See *Hard Target.*

Spoor: A set of tracks or other physical indicators of passage that are visible to a tracker. Spoor is generally interchangeable with "tracks," "set of prints," or "sign."

Spoor Cards: A formal, written sheet on which footprint data are recorded during a follow-up.

Spoor Cutting: Actively searching for spoor; also called "sign cutting" in some organizations.

Spoor Reduction Techniques: *Anti-tracking techniques* in which the quarry splits up, making their spoor progressively more difficult to follow. These techniques include break-away groups, bomb-shelling, and drop-offs.

Spoor Separation Point: A point on the ground where the quarry splits-up into more than one distinct group.

Sticky Messages: Simple, concrete, messages ("touchstones") that have emotional appeal and include compelling storylines.

Straddle: Distance between the inside edge of the left foot to the inside of the right foot, i.e., if the person were standing still with their feet close together, the straddle is the distance between the two feet at their closest points.

Stride: The distance from one footprint to the next in the quarry's direction of movement (left foot to right foot).

Substantiating Evidence: Evidence that is inconclusive by itself, but when taken into account with other evidence, helps "build a case." In tracking, substantiating evidence is any spoor or sign that may indicate the passage of the quarry, but which cannot indisputably be linked to the quarry.

Sustained Observation: Expending conscious energy to observe an area or people over time, in order to develop a sense of "normal."

Tacit Knowledge: Knowledge gained through hands-on practical experience that cannot be written down or easily transmitted. See also *Explicit Knowledge.*

Tactical Cunning: One of the five *combat multipliers*, tactical cunning is the art of "getting into the mind of your adversary," anticipating how they view you, and then employing shrewd and crafty ways to out-think and out-adapt the adversary.

Tactical Patience: One of the five *combat multipliers*, tactical patience is the manipulation of the operational tempo in order to obtain the most advantageous situation.

Tactical Operations Center: See *TOC.*

Tactical Shortcut: See *Heuristics.*

Tracking Team Leader: The tracking team leader controls the follow-up and is responsible for its tactical decisions, movement, formations, and the general conduct of the team.

Template: A design or pattern that guides the design or construction of identical items. In other words, a template is an exact specification.

Template Matching: In decision-making, a template match is an exact match of a person, place, or item.

Terminal Point: The last part of the foot to leave the ground, usually the toes.

Third-Order Effects: The reaction to (or effective of) a second-order effect. The tertiary or far downstream outcome of an incident. See *Second-Order Effect.*

Time and Distance Gap: The distance between a combat tracking team and its quarry. The theoretical distance in which the quarry could have moved between the time of the incident and the time in which the trackers began the follow-up.

Time/Shadow Effect: The time of day determines the angle of the sun and, consequently, the length and direction of shadows.

TOC: A unit's command-and-control hub, assisting the commander in synchronizing operations. TOCs act as the primary driver of the intelligence cycle: receiving, analyzing, integrating, and distributing information across distributed teams.

Toe Dig: In tracking, the indentation (*ground spoor*) left by the force of the toes pushing-off of, and leaving, the ground.

Top-Down Processing: The influence of contextual effects on what is perceived. Contextual effects can include emotions, expectations, motivation, culture, and experiences.

Track Line: The continuous line of observable clues (indicators), visible to the tracker, indicating the path of their quarry.

Track Trap: A piece of terrain where spoor can easily be identified.

Tracker: The member of a tracking team who is physically looking for and following a set of tracks.

Tracker 360: A combat tracking *lost spoor procedure*, in which the tracker systematically move around in a circle in an attempt to locate the lost tracks.

Tracker Support Team: This is a group of armed individuals who may accompany a tracking team, always in the rear and in radio contact, to provide additional fire-power to the tracking team if the tactical situation requires it.

Tracking Team: When tracking or conducting a follow-up of armed and dangerous fugitives a five- or six-person team is employed. A tracking team consists of a tracker (or optionally two trackers), two flank trackers, a team leader, and a rear security tracker.

Transference: Dirt/vegetation or other spoor/sign carried from its natural location and deposited elsewhere.

Tunnel Vision: A natural observational limitation. During periods of high stress, people may develop tunnel vision. Physiologically, tunnel vision literally means reduced peripheral vision. The phrase is also used metaphorically to imply that individuals are attending to fewer cues and ignoring important tasks.

Urban Masking: Camouflage used to disguise—or mask—a person, group, object, or activities in an urban (rather than rural) environment. Actively attempting to blend into the baseline of an urban context.

Urban Tracking: The act of conducting a *follow-up* in an urban setting.

Warrior Ethos: Ethos is the fundamental character or disposition of a community, group, or person. Hence, "warrior ethos" is the fundamental spirit, beliefs, customs, and practices of warriors. An idealized warrior model.

Working Memory: Also called "short-term memory," working memory handles the interim processing of incoming information. Information in working memory is stored for only a few seconds, unless it rehearsed, and it can only store about seven plus-or-minus two (7±2) pieces of information at a time.

Y Formation: The standard formation used by a tactical tracking team. It provides the best balance between security, speed, and control. In this formation the flankers are at a 45° angle ahead of the tracker, who is followed by the tracking team leader, and rear security tracker.

Suggestions for Setting Up a Course

This section contains *suggestions* on how a CODIAC course may be organized. These suggestions are based upon the experiences of the authors, and they may (or may not) work for your organizational structure.

Schedule

For trainers organizing a CODIAC course, their first consideration should be to solidify the dates and times during which the course will occur. Coordinate the schedule with all instructors and key range control personnel. Ideally, the course should be delivered as a single training exercise, i.e., approximately 20-days of instruction (plus two rest days during this period). A sample schedule is included on the resource DVD.

Trainees

Trainees should have similar backgrounds and levels of experience. An ideal class would have approximately 20–30 students—small enough to facilitate group discussions. However, classes can be larger, especially if assistant instructors are available to assist with classroom-based group discussions and field exercises.

Billeting and Messing

After a schedule is agreed upon, trainers should next arrange for billeting. It is recommended that all trainees stay in a common billeting area. Additionally, if possible, house instructors in a separate, but nearby, facility.

If billeting on post, arrange with mess hall to provide messing for attendees. If necessary, arrange scheduling with the mess hall, e.g., possible need to open early. During field exercises, trainees will be required to carry their own food; the course should consider providing these supplies, e.g., a bag lunch or MRE. Note, it is recommended that trainees receive at least one hot meal a day, particularly a hot breakfast.

Transportation

If trainees are billeting in common facility, then course organizers should consider using group transportation. Vehicle footprints should be kept to a minimum, such as providing a bus for trainees and a few utility vehicles for instructors and support staff.

Contractors

Additional contractors will likely be required to carry out the training. In particular, role-players and special effects artists may be needed. Many companies can provide these services. Speak with a local contracting office to see what options are available.

Role-player Contractors

To facilitate some of the macro-tracking (Unit 8) and final (Unit 9) scenarios, role-players will be required. Role-players may be recruited from the trainees' Service or Agency (this is discussed in the next section), or contractors may be hired to completely, or partially, fill scenario roles. Ideally, hired role-players will be ex-patriots from the targeted area of operations, although this is not always necessary. However, it is key that the same role-players participate for the entire course; that is, it is important to maintain role continuity for the duration.

Special Effects Contractors

Trainers should consider hiring special effects experts to simulate the effects of weaponry native to the targeted operational environment. This may include simulated IEDs, RPGs, and small arms fire. Special effects help create a memory–emotion link between a critical incident and the training. Thus, trainers are strongly encouraged to include high-quality special effects, when appropriate.

Classroom

A classroom should be reserved for the lecture-based modules and after-action reviews. This facility should have a computer, projectors, and white boards, as well as sufficient room for movement and reconfiguration during group activities.

Spoor Pit

Trainers should identify a suitable "spoor pit" within quick walking distance of the classroom. A spoor pit is a sand- or dirt-covered piece of ground in which footprints (i.e., ground spoor) can be readily observed. A rake is required to smooth-out the spoor pit between demonstrations.

Range Access

Trainers must secure access to ranges for the duration of the exercise. Coordinate with your local Range Control personnel to reserve the range facilities and set-up ranges to the necessary specifications.

Range Access for Macro-Tracking Exercises

For Unit 8, which includes several macro-tracking exercises, trainers should secure access to variety of terrain types. Ideally, each of the different types of terrain locally available should be used for at least one exercise. These terrain types may include:

- Savannah
- Rain forest/jungle
- Mountains
- Piedmont (foot hills)
- Desert/tundra
- Coastal plains
- Urban environments
- Agricultural areas
- Hardwood or coniferous forest

Identify less challenging terrains for use with the initial exercises and then gradually advance to more challenging ones. Additionally, consider the size of the range or park in which the tracking exercises are held. It must be sufficiently large to facilitate exercises without tracking teams "bumping" into each other.

Range Access for CODIAC Final Scenarios

For Unit 9, trainers will require access to at least one "cognitive ville." It is critical to design this range realistically, so that it accurately represents the operational environment of interest. Additionally, buildings in the "cognitive ville" should also include set-dressing in order to provide sufficient cues for the trainees to interpret the scene. The specific placement and decoration of the building on this ville will vary, depending upon the area of operations; however, most traditional range set-ups, such as those designed for room-clearings exercises, will not suffice. If necessary, Range Control personnel can help acquire and position, or reposition, CONEX boxes on the ville.

To completely support the Unit 9 scenarios, trainers should also reserve a secondary range. This range is used for enhanced observation and basic team tactics practice, and it can be used as an alternate site for some of the scenarios.

OPs and TOC

Trainees will need access to Observation Points (OPs) and a Tactical Operations Center (TOC) for the ville-based scenarios. OPs should be positioned at various distances around the ville, the nearest should be approximately 300 meters outside of the edge of the ville and the farthest can be around 1000 meters from the ville. OPs can be natural observation points, such as a raised outcropping, or they may be man-made, such as a CONEX box with camouflage netting draped over it. The TOC should be placed in a convenient location (e.g., near a watering-hole or by an OP). Participants in the TOC should not be able to readily see the ville from their location.

Safety and Support Vehicles

While on the ranges, water and first aid vehicles will likely be required. A central watering location should be designated, and scenarios should be organized so that trainees can arrive at the central "watering hole" at least once during a day's field exercises. Similarly, if available, portable toilet facilities should be positioned at the "watering hole," as well as near Observation Points on the "ville" range.

Communications

Several communication networks will be required to facilitate the field exercises. These include:

- Instructor and support staff network
- Trainee network
- Role-player network

Each trainee team should have its own frequency. For the tracking scenarios, each team need only monitor its own frequency. For the ville-based scenarios, teams should primarily communicate with the TOC, but as appropriate, they should communicate horizontally with each other.

Equipment Checkout

In addition to communications gear, which will likely require course organizers to check-out, trainees will need access to their organic equipment, including weapons and optics, for Units 3, 8, and 9. For Unit 3, trainees only require access to their organic optical devices. However, for Units 8 and 9, trainees should wear their full kits. It is recommended that trainees be responsible for acquiring their own equipment; thus, sufficient time should be allotted to allow equipment check-out.

Knowledge, Skills, and Attitudes

A set of KSAs is introduced at the beginning of this POI. The KSAs specified provide a framework to represent the training content within this course. They should also be referenced when identifying training objectives, as well as when creating standards and conditions of training. All KSAs were designed to be:

- *Observable:* Observers can reasonably agree when they occur by watching trainees' behaviors
- *Measurable*: They can be quantified on some scale, making comparisons among individuals, teams, or specified criteria possible
- *Trainable:* Instructors can, using multiple methods of instruction, support practice of these underlying competencies, reinforce their occurrence, and sustain them once deployed.

Creating Scenarios

Unit 9 provides some insight into creating scenarios. As discussed in that unit, instructors must create unique scenarios that are appropriate for their trainees, the trainees' operational environment, and the current time period.

Write a Narrative

When designing a scenario, begin by outlining a brief narrative of the events. Most (if not all) of the CODIAC scenarios should occur sequentially—they should explicitly build upon one another.

Write the MSEL

Next, outline the Master Sequence of Events List (MSEL), or the chain of discrete events that will occur in the scenario. Each event in the list should be tied to one or more learning objectives, measures of performance, and measures of effectiveness. Example MSELs are provided on the Resource DVD.

Trainee Briefing

Depending upon their service agency, the trainees should receive some sort of mission briefing or five-paragraph order. Prepare this mission brief. In it, specify the rules of engagement, team roles and responsibilities, operational narrative, goals, and available support. Keep in mind, real-world mission briefings do not always contain complete and correct intelligence.

Logistics

Determine the logistical requirements of the scenario. How many role-players are required? What roles are necessary? What costumes or sets are needed? Are any special effects required? Where are specific characters needed at specific times? Who will serve as the "puppet master," directing role-players and their actions from behind the scenes? How many out-of-play instructional facilitators will be needed?

Rehearsal

The role-players will require at least one walk-through rehearsal of each scenario, to ensure they all know their places and scripted behaviors.

Recommended Reading

Both Greg Williams and David Scott-Donelan recommend many books.

Greg Williams' top recommendations include:

1. *The book of five rings* (Miyamoto Musashi)
2. *Blink: The power of thinking without thinking* (Malcolm Gladwell)
3. *The tipping point: How little things can make a big difference* (Malcolm Gladwell)
4. *The gift of fear: Survival signals that protect us from violence* (Gavin De Becker)
5. *The 33 strategies of war* (Robert Greene)
6. *Inside the revolution* (Joel C. Rosenberg)
7. *On war* (General Carl Von Clausewitz)
8. *On killing: The psychological cost of learning to kill in war and society* (Dave Grossman)
9. *On combat: The psychology and physiology of deadly conflict in war and in peace* (Dave Grossman)

David Scott-Donealn's top picks include:

1. *Tracking—Signs of man, signs of hope* (David Diaz)
2. *Manhunting: Reversing the polarity of warfare* (George A. Crawford)
3. *The SAS guide to tracking* (Bob Carss)
4. *Foundations for awareness, signcutting and tracking* (Rob Speiden)
5. *Tracking: A Blueprint for learning how* (Jack Kearney)
6. *Mantracking, Introduction to the Step-By-Step Method* (Roland Robbins)
7. *The SAS tracking & navigation handbook* (Neil Wilson)
8. *Footwear impression evidence: detection, recovery, and examination* (William J. Bodziak)
9. *Terrorist trail: Backtracking the foreign fighter* (H.John Poole)

Measuring Performance

Instructors are encouraged to collect the following two kinds of evaluation.

Reactions

In order to evaluate trainees' perception of and satisfaction with the training, instructors are encouraged to collect periodic course evaluations and/or end-of-course reactions. Reactions surveys, provided on the Resource DVD, may be used for this purpose. Additionally, instructors are encouraged to gather feedback from the trainees during AAR, which should be carried out periodically during curriculum delivery.

Learning

Knowledge and skills acquired and/or improved during training should also be measured through traditional testing and observation during practical exercises.

Knowledge acquisition should be assessed by using pre- and posttests; versions of such tests provided on the Resource DVD. Skill acquisition should be evaluated during the practical exercises via observation. Behavioral Observation Checklists (BOCs) are provided on the Resource DVD to assist with this activity.

Descriptions of Scenario Clips (Unit 9)

This section includes brief descriptions of the scenario video clips provided on the Resource DVD.

9001.1 – AAR – Greg Williams

In this clip, Greg Williams walks trainees through the "Pulling Teeth" scenario's after-action review. He discusses the importance of using clear and concise communication, thinking about information priority, and communicating with the TOC. He also discusses techniques for recording intelligence in the TOC, as well as better ways to employ team tactics (from the OPs) to conduct observation. Finally, Williams reminds trainees to watch their "7, 6, 5, and 3," which refers to the seven-step terrorist planning cycle, six domains of combat profiling, five combat multipliers, and rule of three.

9001.2 – AAR – Will Atkinson

This clip is a continuation of video 9001.1. In this video, Will Atkinson walks trainees through the specific narrative that they observed in Scenario #1.

9002.1 – Scenario 02 – Synopsis

This video shows a synopsis of the whole Scenario #2. At the beginning of the video, one of the Red Crescent vehicles becomes separated from the convoy; it drives up to the marketplace to ask for directions. Next, the insurgents are seen. They enter the marketplace and, generally, cause a ruckus. Meanwhile, the lead insurgent (wearing the green headscarf and "Miami Ink" black shirt) gives a signal by dropping his bag. As Prince Habibi brings in the police to deal with the ruckus in the market, shots ring out and chaos erupts. The audio heard in this video is from the "puppet master's" radio.

9002.1 - Scenario 02 – RPG

This short clip shows the actual RPG attack on the lost Red Crescent vehicle. The video has no audio.

9003.1 – Scenario 03 – Insurgents Perspective

This video shows the insurgents practicing their tactics, building explosive devices, and then detonating a suicide bomb at the Red Crescent's location. The video has no audio.

9003.2 – Scenario 03 – Synopsis

This video shows a synopsis of the Scenario #3. The day appears to be progressing like any other, then at the end, the insurgents strike—detonating their explosives and opening fire on the police. The audio heard in this video is from the "puppet master's" radio. Most of the audio is direction to the role-players, instructors, or videographers; however, some of the audio is on the "626" radio channel—this is a communications channel that the TOC can hear in which the US Command Element engages in dialog with their embedded local contact.

9004.1 – Scenario 04 - Police Assassination

This clip shows the staged assassination of one of the police officers, and then the return of his body to the villagers. The video includes no audio.

9005.1 – Scenario 05 – Synopsis

This video shows a synopsis of Scenario #5; it is shot in night vision. The villagers are holding a funeral for the victims of the recent violence. The video opens with the villagers slowly gathering, then it cuts to a clip of the trainees observing the village with their night-vision optics. Next, one of the insurgents can be seen trying to sneak away. He is caught acting suspicious, so he tries to blend into the funeral crowd, but with the trainees' help he is apprehended by the police.

9006.1 – Scenario 06 – Synopsis

At the beginning of the video, Prince Habibi can be seen meeting with his extra security contingent. They travel together to where the Imam is holding a "mini *shura*" with villagers; then the villagers can be seen putting dead bodies in the back of a truck and mourning their dead. Shots ring out at the funeral, but these are fired in grief (up in the air). The video then cuts to an interview with Prince Habibi, who is talking about his new security regiment. (The video crew is playing the part of a news crew.) The police have just apprehended one of the terrorists; Habibi is pleased but cautious. The video then cuts to a scene where Habibi and the Imam are meeting with the Bedouin leaders. Habibi is trying to recruit more people to fight against the insurgents.

9006.2 – Scenario 06 – Assassination of the Bedouin Leader

This video, shot from the terrorists' perspective, shows the assassination of the Bedouin leader, in retaliation for his support of Prince Habibi. The video includes no audio.

9007.1 – Scenario 07 – Synopsis

The beginning of the video shows Prince Habibi, then the news crew is interviewing townsfolk. Next, one of the villagers is seen carrying a white flag and heading off to the talk with the insurgents; he is sent back to the village beaten (and wearing only his underwear). The Prince, Imam, and their crew escort him to the Prince's home. Soon thereafter, the insurgents arrive in force and hold a *shura* with the Prince and Imam. The man wearing the red hat is a figurehead, while the insurgent in the green headgear is the true insurgent leader. They leave the meeting pretending to agree to peace. As the Prince is discussing this with the camera crew, the insurgents attack the mosque. The Imam is killed in the attack; the Prince is overcome by grief and begins shooting wildly towards the insurgent compound. His people try to intervene and protect him. Finally, the video ends with a short after-action review by Greg Williams, praising the role-players' performance.

9007.1 – Scenario 07 – Beating

This clip shows the insurgents beating the Prince's envoy who originally asked for peace. It includes no audio.

9008.1 – Scenario 07 – Synopsis

This video shows one team meeting with the village Sheik, Prince Habibi, and his retinue. The people wearing red shirts (specifically Greg Williams) are out-of-play—they are acting as instructor/observers. The Soldiers can be seen attempting to gain the trust of the prince; although they are clearly struggling, they are attempting to engage in proper cultural behaviors and build trust with the Prince.

9009.1 – Scenario 07 - Synopsis

This short clip, shot in night vision, shows insurgents planning an IEDs on the road. Later, as the route clearing team attempts to deal with the IED, they are attacked by an insurgent with an RPG. This clip has no audio.

Printed in Great Britain
by Amazon